FOUNDATIONS TODAY SERIES
2005 EDITION

FOUNDATION YEARBOOK

Facts and Figures on Private and Community Foundations

Steven Lawrence
Director of Research

Josefina Atienza
Research Analyst/
Coordinator of the *Foundations Today Series*

Asmita Barve
Research Associate

THE
FOUNDATION
CENTER

CONTRIBUTING STAFF

Loren Renz	Vice President for Research
Kathye Giesler	Publishing Database Administrator
Emmy So	Database Operations Assistant
Cheryl L. Loe	Director of Communications
Christine Innamorato	Production Coordinator, Publications

ABOUT THE REPORT

The original research upon which this report is based was conducted by the Foundation Center. Data from the report may not be cited or reproduced without attribution to *Foundation Yearbook* and the Foundation Center. Other reports in the *Foundations Today Series* of annual research studies include *Foundation Giving Trends* and *Foundation Growth and Giving Estimates*. In addition, the Center produces numerous studies on special topics including:

- *Foundation Funding for Children's Health: An Overview of Recent Trends*, 2005
- *Key Facts on Family Foundations*, 2005
- *Giving in the Aftermath of 9/11: Final Update on the Foundation and Corporate Response*, 2004
- *International Grantmaking III: An Update on U.S. Foundation Trends*, 2004
- *Foundation Funding for the Humanities: An Overview of Current and Historical Trends*, 2004
- *Update on Foundation Health Policy Grantmaking*, 2004
- *9/11 Relief and Regranting Funds: A Summary Report on Funds Raised and Assistance Provided*, 2003
- *Update on Funding for Higher and Graduate Educational Institutions*, 2003
- *Arts Funding IV: An Update on Foundation Trends*, 2003
- *New York Metropolitan Area Foundations: A Profile of the Grantmaking Community*, 2002

ABOUT THE FOUNDATION CENTER

The Foundation Center's mission is to strengthen the nonprofit sector by advancing knowledge about U.S. philanthropy.

To achieve our mission, we:

- Collect, organize, and communicate information on U.S. philanthropy
- Conduct and facilitate research on trends in the field
- Provide education and training on the grantseeking process
- Ensure public access to information and services through our Web site, print and electronic publications, five library/learning centers, and a national network of Cooperating Collections.

Founded in 1956, the Center is the nation's leading authority on philanthropy and is dedicated to serving grantseekers, grantmakers, researchers, policymakers, the media, and the general public.

ACKNOWLEDGMENTS

The authors gratefully acknowledge the many Foundation Center staff whose work provides the information base necessary for this analysis. We also thank the many foundations that report their grants directly to the center, providing more detailed information not otherwise available.

ISBN 1-59542-017-7   ISSN 1539-7629

This report is available for $45 or through subscription to the annual *Foundations Today Series* of research reports. Subscriptions are $95.00 each and include three reports. Discounts are available on bulk orders or subscriptions. To order by phone or for more information, call toll-free (800) 424-9836.

Contents

Tables and Figures

Additional Resources

Visit FC Stats, the Foundation Center's free online statistical information service, at www.fdncenter.org/fc_stats for additional current and historical data tables on the universe of U.S. grantmaking foundations. Resources include:

- Historical data on number of foundations and foundation giving, assets, and gifts received
- Aggregate financial data by foundation type
- Lists of top funders by foundation type
- Aggregate financial data by state and for selected metro areas
- List of top funders by state
- Staffing data
- Public reporting data

CHAPTER 3: Foundations by Region and State

CHAPTER 4: Foundations by Foundation Type

Foreword

Through numerous research reports published over the past two decades, the Foundation Center has created for the field an increasingly refined and detailed portrait of foundation philanthropy. This rich collection of facts and figures on private and community foundations and their grantmaking interests begins with 1975 and continues through over 25 years of foundation development. In the 1990s, the principal vehicle for presenting and interpreting this growing body of data on foundation growth and giving trends was *Foundation Giving,* an annual statistical yearbook. From 1991 to 1999, *Foundation Giving* doubled in size, paralleling the meteoric growth of the foundation field, and demonstrating the Center's continuous efforts to enhance and expand its examination of foundations.

In 2000, we introduced the *Foundations Today Series,* the successor to *Foundation Giving.* The *Foundations Today Series* organizes foundation trends into three annual reports on the current state of foundations and their giving. The series approach allows us to present information in a more timely and tailored way. The first report of the 2005 series, *Foundation Giving Trends,* presents a picture of how just over 1,000 of the top U.S. foundations distributed their grant dollars in 2003. *Foundation Growth and Giving Estimates* provides highly informed predictions of giving levels in 2004 and suggests directions for 2005. Finally, *Foundation Yearbook* tracks foundation growth and development for over 66,000 foundations active in 2003.

While concentrating on trends of the most recent year, reports in the *Foundations Today Series* also compare growth and giving patterns retrospectively. The series is the only source providing a comprehensive record on foundation fiscal data; new foundation development; and changes in levels of support by grant subject or purpose, recipient type, type of support, and population group.

Recognizing that foundations are an increasingly diverse community, the *Foundations Today Series* has been designed to satisfy information needs concerning, first, the universe of over 66,000 grantmaking foundations, and, second, specific segments of that universe. A wealth of summary data can be found for independent, corporate, community, and grantmaking operating foundations. Foundation statistics are further segmented regionally and by state. Finally, the behavior of large and small foundations and the patterns of large and small grants are independently evaluated.

Long-time readers of *Foundation Yearbook* may notice changes in the format and content of the 2005 edition. This new look reflects our efforts to streamline the *Foundation Today Series* and make the reports more user-friendly. We welcome comments about these changes, as well as other suggestions to enhance or improve the series.

Loren Renz
Vice President for Research

Key Findings

The beginning of the stock market recovery in 2003 and a higher level of new gifts into foundations helped revive assets and giving by the nation's more than 66,000 grantmaking foundations in the latest year. Overall, foundation giving grew an estimated 6.9 percent to $32.4 billion in 2004, up from $30.3 billion in 2003. This followed a 0.4 percent dip in giving in the prior year. Foundation assets increased 9.5 percent in 2003, ending two consecutive years of decline. Nonetheless, assets remained below the peak levels reached in 2000. By foundation type, community foundations increased their giving fastest in 2004, followed by independent and corporate foundations.

Helping to boost giving was the continued (albeit slower) growth in the number of active grantmaking foundations. Foundations in the West posted the fastest growth in number in 2003. Western foundations also surpassed Midwestern foundations by share of overall giving—for the first time on record. Reflecting the improved economic climate, foundation assets rose in all four major regions, with the West posting the largest gain.

Foundation Yearbook, 2005 Edition, documents changes in the actual number, giving, and assets of all active U.S. foundations through 2003 and provides estimates of foundation giving through 2004. The report provides an overview of the state of foundation giving in the current year and beyond; comparisons of foundation activities by foundation size; and breakdowns of foundation resources by geographic location and grantmaker type. *Foundation Yearbook* is part of the *Foundations Today Series* of annual research reports on foundation growth and giving.

Findings presented in *Foundation Yearbook* are based on aggregate fiscal data collected by the Foundation Center on all active U.S. foundations. Estimates for 2004 are based on survey figures reported by over 900 large and mid-size foundations combined with year-end fiscal indicators.

State of Foundation Giving

- Giving rebounded to an estimated $32.4 billion in 2004, up 7 percent from $30.3 billion in 2003

- Giving in 2005 is likely to increase modestly, with growth tempered by economic uncertainty and fluctuations in the stock market

Foundation Growth, Giving, and Assets Through 2003

- Number of foundations reached a record 66,398 in 2003, up 2.4 percent from 2002

The number of states with at least 1,000 foundations jumped from seven in 1975 to 22 in 2003

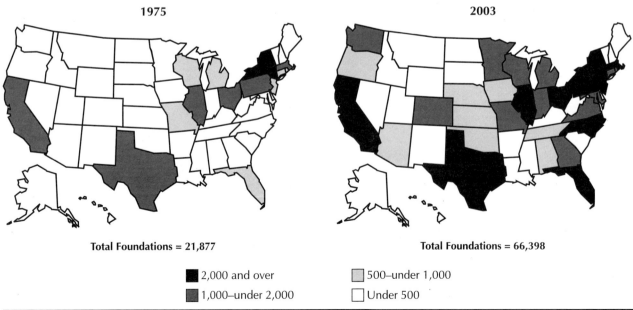

1975

2003

Total Foundations = 21,877

Total Foundations = 66,398

- **2,000 and over**
- **1,000–under 2,000**
- 500–under 1,000
- Under 500

Source: The Foundation Center, *Foundation Yearbook*, 2005.

- Number of grantmaking foundations increased by over 1,500 in the latest year—compared to a peak gain of nearly 6,400 in 2000

- Giving by the nation's grantmaking foundations slipped 0.4 percent to $30.3 billion in 2003

- Giving by the 25 largest funders rose 10 percent in 2003 to $6.6 billion

- Assets of all active U.S. foundations rebounded to $476.7 billion, but remained below the record $486.1 billion reached in 2000

- Assets of the 25 largest foundations climbed 17 percent in the latest year, well above the 9.5 percent rise in overall foundation assets

- Gifts into foundations in 2003 totaled $24.9 billion, a 12 percent gain over the prior year

- Total value of gifts of $5 million or more grew 18 percent to $17.9 billion

Foundations by Region and State

Regions:

- West experienced fastest rate of growth in number of foundations in 2003

- South reported largest gain in actual number of foundations since 1975

- Northeast ranked first by percentage growth in giving in 2003

- West surpassed Midwest by share of overall giving for the first time on record

- All four regions showed asset gains in latest year

- West registered largest increase in assets in 2003

Foundations in five states accounted for close to half of 2003 giving

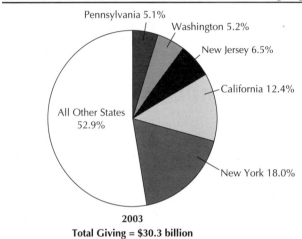

2003
Total Giving = $30.3 billion

Source: The Foundation Center, *Foundation Yearbook,* 2005.

Despite slight reductions in 2002 and 2003, foundation giving has more than doubled since 1997

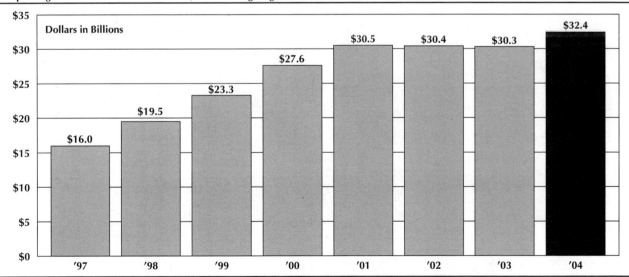

Source: The Foundation Center, *Foundation Yearbook,* 2005. Figure estimated for 2004. All figures based on current dollars.

States:

- Maine experienced most rapid rate of growth in giving in 2003

- New Jersey led in actual grant dollar gain in latest year

- New York foundations reported largest amount of giving overall

- District of Columbia ranked first based on giving per capita due to its small population and the presence of a few large national and international funders

- Delaware registered fastest rate of growth in assets in 2003

- California topped all states in actual asset dollar gain

- All but four states reported asset increases in 2003

Giving and Growth of Independent, Corporate, and Community Foundations

- Giving by independent foundations decreased 2.9 percent in 2003, the second consecutive year of decline

For the first time on record, the West surpassed the Midwest by share of overall giving in 2003

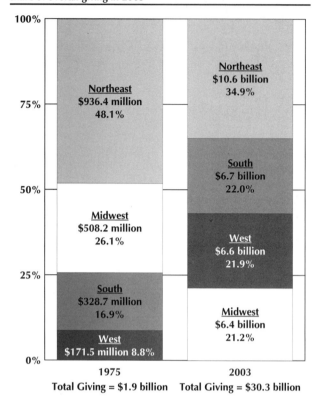

1975	2003
Total Giving = $1.9 billion	Total Giving = $30.3 billion

Source: The Foundation Center, *Foundation Yearbook,* 2005. Figures based on unadjusted dollars. Due to rounding, percentages may not total 100. Total giving figure includes foundations based in the Caribbean and South Pacific, which accounted for 0.0 percent of total foundation giving in 1975 and 2003.

Maine led all other states in the rate of growth of foundation giving between 2002 and 2003

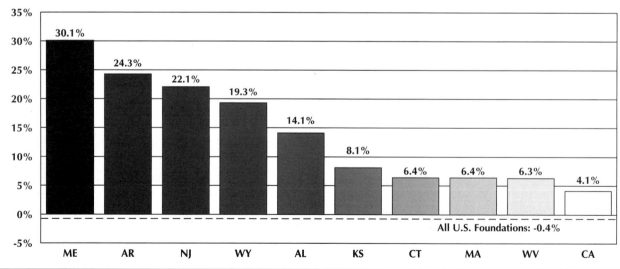

Source: The Foundation Center, *Foundation Yearbook,* 2005. Percent change based on current dollars.

- Independent foundation assets rose 9.6 percent, following a 7.6 percent decline in 2002

- Corporate foundation giving inched up 0.2 percent in 2003, compared to a 5.3 percent gain in 2002

- Assets of corporate foundations rebounded to $15.4 billion, a 7.1 percent gain from 2002

- Giving by community foundations registered a 0.2 percent uptick in 2003, following a 5.1 percent rise in 2002

- Community foundation assets increased 14.7 percent in the latest year, exceeding the asset gains reported by independent, corporate, and grantmaking operating foundations

Private (non-operating) and independent foundation giving in 2004 totaled 6.7 percent of the prior year's assets

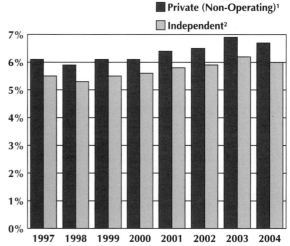

Source: The Foundation Center, *Foundation Yearbook*, 2005. Giving figures for 2004 based on estimates.
[1] Figures exclude private foundations that have been defined by the IRS as "Operating" foundations. These foundations are not subject to the same payout requirement as other private foundations. Community foundations are excluded as they are grantmaking public charities, not private foundations.
[2] Figures exclude private foundations identified by the Foundation Center as serving as vehicles for corporate philanthropy, many of which operate as "pass-through" foundations.

Foundation Profile, 2002 to 2003

	2002	2003	% Change
All Foundations			
No. of Foundations:	64,843	66,398	2.4
Total Giving:	$ 30,432	$ 30,309	-0.4
Total Assets:	$435,190	$476,713	9.5
Gifts Received:	$ 22,163	$ 24,858	12.2
Independent			
No. of Foundations:	57,834	58,991	2.0
Total Giving:	$ 23,254	$ 22,568	-2.9
Total Assets:	$364,143	$399,138	9.6
Gifts Received:	$ 13,952	$ 15,846	13.6
Corporate			
No. of Foundations:	2,362	2,549	7.9
Total Giving:	$ 3,457	$ 3,466	0.2
Total Assets:	$ 14,428	$ 15,447	7.1
Gifts Received:	$ 3,002	$ 3,234	7.7
Community			
No. of Foundations:	661	699	5.7
Total Giving:	$ 2,526	$ 2,532	0.2
Total Assets:	$ 29,772	$ 34,153	14.7
Gifts Received:	$ 3,175	$ 3,476	9.5
Operating			
No. of Foundations:	3,986	4,159	4.3
Total Giving:	$ 1,177	$ 1,744	46.0
Total Assets:	$ 26,847	$ 27,975	4.2
Gifts Received:	$ 2,035	$ 2,302	13.1

Source: The Foundation Center, *Foundation Yearbook*, 2005. Dollars in millions. Percent change represents current dollars. Includes only foundations that awarded grants in the latest fiscal year.

Only two percent of foundations held nearly seven out of ten asset dollars in 2003

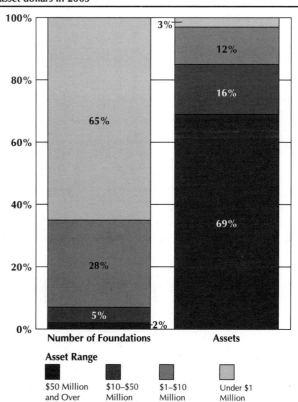

Source: The Foundation Center, *Foundation Yearbook*, 2005.

CHAPTER 1

The State of Foundation Giving, 2005

- **Foundation giving rose 7 percent to an estimated $32.4 billion in 2004**

- **2004 assets increased a more modest 4 to 6 percent**

- **Growth in 2005 giving likely to be modest**

At the end of 2003, it appeared that foundations were poised to return to a period of steady, substantial increases in giving. The stock market posted a double-digit gain, following three consecutive years of declines, and foundation assets began to recover. The year-end optimism proved short-lived, however, and the market fluctuated widely in 2004. Despite some positive economic indicators, each worrisome development—an expanding deficit, rising oil prices, a deteriorating war situation, the general uncertainty in the run-up to the presidential election—sent the market down.

In this climate of uncertainty, overall private giving by individuals, foundations, and corporations grew a modest 5 percent to an estimated $248.52 billion in 2004.[1] For the more than 66,000 U.S. grantmaking foundations, estimated giving increased 6.9 percent to a record $32.4 billion (Figure 1). While this gain in foundation giving fell below the 9.5 percent rise in foundation assets in 2003, giving was just over 6 percent higher than in 2001—the previous peak year for foundation giving.

The increase in foundation giving in 2004 resulted primarily from the stock market recovery that began in 2003 and a higher level of new gifts into foundations. An additional factor contributing to the overall growth of foundation assets and giving was a modest increase in newly established foundations. These factors helped to reverse two years of asset losses, although total foundation assets remained below the

FIGURE 1. Change in Foundation Giving, 1994 to 2004

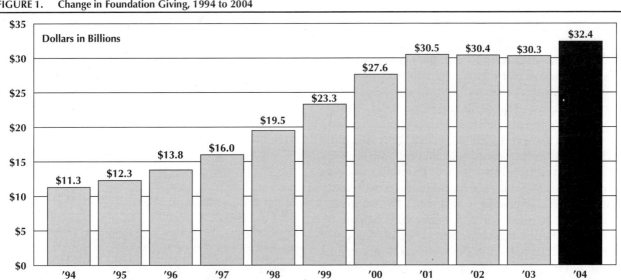

Source: The Foundation Center, *Foundation Yearbook,* 2005. Figure estimated for 2004. All figures based on current dollars.

1

record levels reached in 2000. In fact, assets of most of the largest U.S. foundations had yet to return to their previous peak levels.

Independent foundations, including most family and "new health foundations" (formed from health care conversions), increased their giving by an estimated 6.8 percent in 2004. This followed a nearly 5 percent reduction in giving between 2001 and 2003.[2] Overall, independent foundations gave a record $24.1 billion, which surpassed the previous record of $23.7 billion reported in 2001. (For more detailed comparisons of estimated giving by foundation type, see Chapter 4.)

Corporate foundation giving grew by an estimated 4 percent in 2004, following a 0.2 percent uptick in giving in 2003. Actual giving increased from under $3.5 billion to a record $3.6 billion. Increases in the value of existing corporate foundation assets and a higher level of new gifts into foundations contributed to this growth.

Community foundations increased their giving by an estimated 8.6 percent in 2004, surpassing both independent and corporate foundations. By comparison, community foundation giving grew just 1.4 percent in the prior two-year period. Actual grant dollars totaled an estimated $2.7 billion, up from $2.5 billion in 2003.

Foundation giving rose by 30 percent between 1999 and 2003, despite two interim years of slight overall decreases. Even adjusted for inflation, giving by foundations grew 17.7 percent during this five-year period. Inflation-adjusted foundation giving per capita grew by almost 13 percent, from $74.35 in 1999 to $83.94 in 2003 (Figure 2). Foundation giving as a percent of Gross Domestic Product (GDP) also increased over the same period, from 0.25 percent to 0.28 percent (Figure 3).[3] Nonetheless, inflation-adjusted 2003 figures for overall foundation giving and for giving per capita and as a share of gross domestic product remained below the peak levels recorded in 2001.

In the latter half of the 1990s, the soaring stock market and robust economy and the amount of new gifts and bequests from donors to their existing foundations were the key drivers for the increased value of foundation assets. The rapid rise in personal wealth during this period also led many individuals to create foundations. Between 2000 and 2002, however, the stock market decline and a sluggish economy caused a 10.5 percent drop in the value of foundation assets overall. (Many of the largest foundations experienced a much larger decrease in their assets.) The return of positive stock market performance in 2003 helped to reverse this trend, although the shakier market performance in 2004 will contribute only modestly to the rebound in foundation giving.[4]

An additional factor slowing the recovery in foundation giving is the practice of many permanently endowed foundations—including most of the nation's largest ones—of following long-term payout strategies. These strategies helped to mitigate reductions in foundation giving through 2003 but are also contributing to slower rises in giving during the early years of the economic recovery. For example, Figure 4 illustrates the lag between the decline in inflation-adjusted foundation assets in 2001 and the subsequent decrease in real-dollar giving, which did not occur until 2002. Conversely, despite the return to growth in foundation assets in 2003, giving continued to decrease. (In contrast, individuals' personal giving usually mirrors fairly directly real or anticipated changes in their net worth.)

The federal payout requirement for private, non-operating foundations also helps to stabilize foundation giving in both good and bad times. By law, independent and corporate foundations must pay out each

FIGURE 2.   Foundation Giving Per Capita, 1994 to 2003 (Constant Dollars)

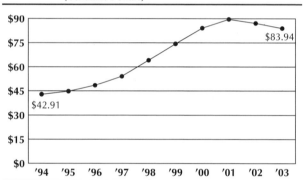

Source: The Foundation Center, *Foundation Yearbook,* 2005. Dollars in billions. Constant 1994 dollars based on annual average Consumer Price Index, all urban consumers. Figures on U.S. population are based on resident population figures. See U.S. Census Bureau, *Statistical Abstract of the United States: 2004–2005* (124th Edition), Washington, DC, 2004.

FIGURE 3. Foundation Giving as a Share of GDP, 1994 to 2003

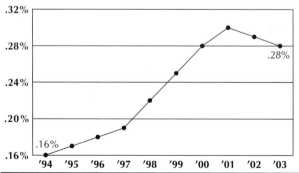

Source: The Foundation Center, *Foundation Yearbook,* 2005. Figures on U.S. Gross Domestic Product (GDP) are from the U.S. Bureau of Economic Analysis. See U.S. Census Bureau, *Statistical Abstract of the United States: 2004–2005* (124th Edition), Washington, DC, 2004.

year in charitable distributions at least 5 percent of the value of their assets in the preceding year. (They may carry forward payout in excess of 5 percent over several years.) Figure 5 shows that estimated independent foundation giving in 2004 was 6 percent of the prior year's assets—the second highest ratio recorded since the Foundation Center began separate tracking of independent foundations in 1987. Smaller independent foundations generally exceed this rate because many of them maintain minimal endowments and instead serve as "pass-throughs" for charitable giving by their donors. For larger, endowed independent foundations, the relatively high levels of payout seen during the economic downturn and continuing into the start of the recovery may reflect a desire to maintain stable giving levels despite variations in asset values.

FIGURE 4. Change in Foundation Assets and Giving, 1975 to 2003 (Constant Dollars)

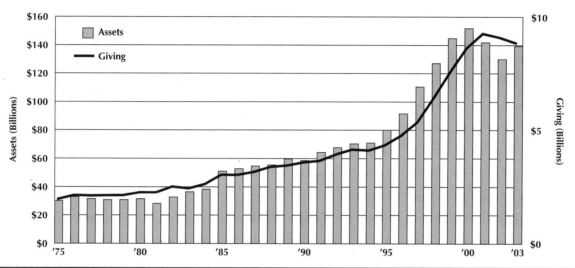

Source: The Foundation Center, *Foundation Yearbook,* 2005. Constant 1975 dollars based on annual average Consumer Price Index, all urban consumers, as reported by The U.S. Department of Labor, Bureau of Labor Statistics, as of February 2005.

FIGURE 5. Private (Non-Operating) and Independent Foundation Giving as a Share of Prior Year's Assets, 1988 to 2004

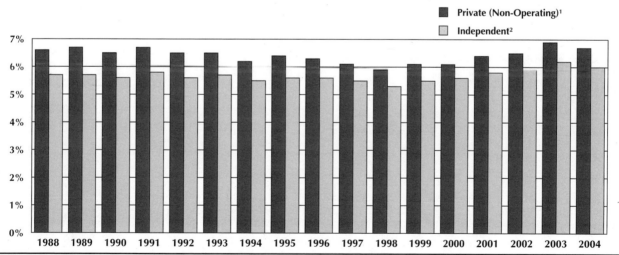

Source: The Foundation Center, *Foundation Yearbook,* 2005. Giving figures for 2004 based on estimates.

[1]Figures exclude private foundations that have been defined by the IRS as "Operating" foundations. These foundations are not subject to the same payout requirement as other private foundations. Community foundations are excluded as they are grantmaking public charities, not private foundations.

[2]Figures exclude private foundations identified by the Foundation Center as serving as vehicles for corporate philanthropy, many of which operate as "pass-through" foundations.

Note: Private (non-operating) foundations are required to pay out each year at least 5 percent of the value of their investments in the preceding year. (They may carry forward payout in excess of 5 percent over several years.) "Qualifying Distributions" is the amount used in calculating the required 5 percent payout and includes total giving, as well as reasonable administrative expenses, set-asides, PRIs, operating program expenses, and the amount paid to acquire assets used directly for charitable purposes. Actual payout can only be calculated for individual foundations. The ratio of total giving to prior year asset values illustrated above therefore serves as only a rough proxy for payout.

For corporate foundations, payout was 23.3 percent in 2004, nearly unchanged from 2003. Unlike larger independent foundations, corporate foundations generally do not maintain substantial endowments and instead fund grant budgets through annual gifts from their corporate donors—hence, the ratio of giving to the prior year's assets is higher. Many corporations make additional gifts into their foundations in periods of stronger earnings and draw down these assets during periods of weaker earnings.

In contrast to independent and corporate foundations, community foundations, as public charities, have no payout requirement. Nonetheless, their giving typically amounts to between 7 and 8 percent of the value of their prior year's assets.

Figure 6 illustrates the changes in foundation giving in inflation-adjusted dollars since 1975 (the first year for which comprehensive information is available). In the recessions of 1980, 1981-82, 1990-91, and 2001, foundation giving in real dollars did not decline and, in fact, increased slightly. However, giving decreased for two consecutive years following the end of the latest recession.

To assess the outlook for foundation giving in 2005, the Foundation Center's annual "Foundation Giving Forecast Survey" asked respondents to indicate expected changes in their giving in the current year. Responding to the relatively improved economic climate, over half of foundations (55 percent) indicated that their giving would increase in 2005, while one-fifth (20 percent) expected no change in their level of giving (Figure 7).[5] Among foundations expecting an increase, the largest share (29.9 percent) predicted

increases in the range of 1 to 5 percent. An additional 20.7 percent expected to increase giving between 5 and 10 percent. However, one-quarter of respondents (25 percent) expected to decrease their giving in 2005, surpassing the 18 percent of respondents to the 2004 survey that expected to decrease their giving. These findings suggest that foundation giving will likely continue to increase in 2005 but more modestly than in 2004. (See *Foundation Growth and Giving Estimates: 2004 Preview* for a more detailed analysis of the 2005 survey findings.)

FIGURE 7.   Anticipated Changes in 2005 Foundation Giving

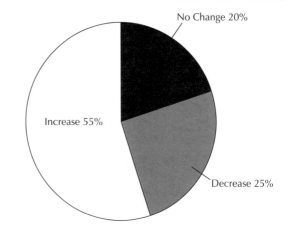

No. of Respondents = 891

Source: The Foundation Center, *Foundation Yearbook,* 2005. Figures based on the Foundation Center's 2004 "Foundation Giving Forecast Survey." For more information on the survey sampling base, see "Methodology" in Appendix A.

FIGURE 6.   Total Giving by Foundations, 1975 to 2004 (Constant Dollars)

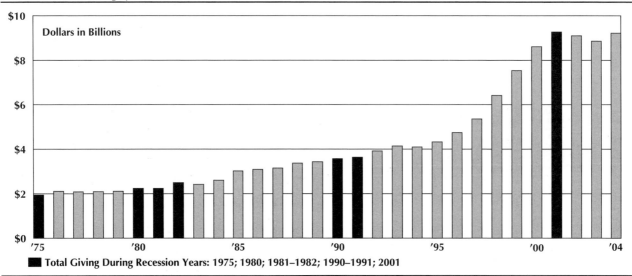

Source: The Foundation Center, *Foundation Yearbook,* 2005. Figure estimated for 2004. Constant 1975 dollars based on annual average Consumer Price Index, all urban consumers, as reported by U.S. Department of Labor, Bureau of Labor Statistics, as of February 2005.

Helping to boost foundation giving in both the short and long term will be the new and sometimes very large foundations established in the past several years and the transfer of exceptionally large gifts and bequests from donors to their existing foundations. While information on foundation creation between 2000 and 2003 remains incomplete (newly established foundations have up to three years to begin grantmaking), the number of larger foundations established during the 1990s has far exceeded the number established in the 1980s, the decade with the highest birth rate up until then. New formation slowed in the sluggish economic climate of the early 2000s. Nonetheless, the number of larger foundations established since the turn of the century has already exceeded the number of active larger foundations formed during any decade prior to the 1980s. As many of these new foundations receive additional assets over the next ten to twenty years, they will contribute substantial resources to the nation's nonprofit community.

Another factor helping to boost the recent growth of foundations was the addition of a permanent provision to the federal tax code in 1998 permitting living donors to deduct the current market value of gifts of appreciated property to private foundations. (Temporary provisions had been in effect throughout much of the 1990s.) By comparison, the Congressional Joint Committee on Taxation proposed several regulatory changes affecting foundations in 2005, including one that would limit the tax incentives for gifts to foundations of real property, closely held business interests, restricted stock, and oil and gas interests. While the likely impact of this proposal on giving remains unstudied, it could have the effect of reducing new gifts into foundations.

Alternatively, reductions in personal income and estate tax rates could slow the establishment of new foundations and reduce the growth of existing foundations through gifts or bequests. As evidenced by the decline in the rate of foundation formation immediately following the 1986 personal income tax reductions, decreases in taxes for wealthy individuals lower the incentive for charitable giving.

Foundations appear to have returned to a path of steady, albeit modest, annual growth in giving. The number of grantmaking foundations also continues to grow, although at a markedly slower pace when compared to the latter years of the late 1990s and the first years of the new century. Nonetheless, baring a dramatic return to the relative political stability and widespread economic prosperity of the prior decade, it seems unlikely that foundations will again realize consistent double-digit gains in their giving in the foreseeable future.

June 2005

Endnotes

1. See Giving USA Foundation, *Giving USA 2005: The Annual Report on Philanthropy for the Year 2004,* ed. by M. Brown, Indianapolis, IN: Giving USA Foundation, 2005. Giving by independent, corporate, community, and operating foundations totaled 13 percent of all private giving in 2004.
2. Percentages based on the Foundation Center's 2005 "Foundation Giving Forecast Survey." A total of 945 surveyed foundations provided estimates of changes in their 2004 giving. For more information on the survey sampling base, see L. Renz and S. Lawrence, *Foundation Growth and Giving Estimates: 2004 Preview,* New York: Foundation Center, 2005.
3. Between 1999 and 2003, private charitable giving from all sources—including individuals, corporations, and foundations—decreased slightly as a share of Gross Domestic Product from 2.3 percent to 2.2 percent. For additional information, see *Giving USA 2005.*
4. Based on the Foundation Center's 2005 "Foundation Giving Forecast Survey," foundation assets grew by an estimated 4 to 6 percent between 2003 and 2004. A total of 912 surveyed foundations provided estimates of changes in their 2004 assets.
5. In general, the largest U.S. foundations are more likely to have endowments that include major equities holdings. As a result, they are more likely to have lost asset value in the bear market of the early 2000s and to have gained value in the stock market recovery that began in 2003.

CHAPTER 2

Foundation Giving Through 2003

Overview

- Number of foundations surpassed 66,000

- Foundation giving decreased slightly to $30.3 billion

- Foundation assets rebounded to $476.7 billion

- Gifts to foundations from their donors increased to $24.9 billion

Foundation giving continued to decline in 2003, despite gains in foundation assets resulting from the beginning of a stock market recovery. Giving slipped 0.4 percent to $30.3 billion, while foundation assets rose 9.5 percent to $476.7 billion. Nonetheless, foundation assets remained below their peak in 2000. The slight decline in giving was partially mitigated by a modest increase in the number of active grantmaking foundations, which reached a record high of 66,398.

2003 Giving

- Foundation giving totaled $30.3 billion, down $123 million from 2002

- For the first time on record, West surpassed Midwest by share of overall giving

Foundation giving totaled $30.3 billion in 2003, down $123 million (0.4 percent) from the prior year (Tables 1 and 2).[1] Adjusted for inflation, the decrease was 2.6 percent. This marked the second consecutive year of marginal declines. Nonetheless, actual grant dollars stood at the third highest level on record. Since 1995, foundation giving has jumped $18 billion, representing an increase of 104.7 percent after inflation (Figure 8). This robust performance has increased foundations' share of all private giving to nearly 13 percent in 2003, up from less than 9 percent in the mid-1990s (see "Foundations' Share of Private Philanthropy, 2003").

TABLE 1.   Aggregate Fiscal Data by Foundation Type, 2003

Foundation Type	Number of Foundations	%	Assets	%	Gifts Received	%	Qualifying Distributions[1]	%	Total Giving[2]	%	PRIs/Loans[3]	%
Independent	58,991	88.8	$399,138,435	83.7	$15,845,858	63.7	$24,523,592	72.3	$22,567,751	74.5	$165,024	86.7
Corporate	2,549	3.8	15,446,861	3.2	3,234,060	13.0	3,895,448	11.5	3,465,605	11.4	4,345	2.3
Community	699	1.1	34,152,980	7.2	3,476,316	14.0	2,532,279	7.5	2,531,530	8.4	10,069	5.3
Operating	4,159	6.3	27,974,838	5.9	2,302,092	9.3	2,989,780	8.8	1,743,950	5.8	10,938	5.7
Total	66,398	100.0	$476,713,115	100.0	$24,858,326	100.0	$33,941,098	100.0	$30,308,835	100.0	$190,376	100.0

Source: The Foundation Center, *Foundation Yearbook,* 2005. Due to rounding, figures may not add up. Dollars in thousands.

[1]Qualifying distributions are the expenditures used in calculating the required 5 percent payout for private foundations; includes total giving, as well as reasonable administrative expenses, set-asides, PRIs, operating program expenses, and the amount paid to acquire assets used directly for charitable purposes.

[2]Includes grants, scholarships, and employee matching gifts; excludes set-asides, loans, PRIs, and program expenses.

[3]Program-Related Investments (PRIs) include low- or no-interest loans and charitable investments for projects clearly related to the foundations' grantmaking interests. These disbursements count toward qualifying distributions.

The decline in foundation giving in 2003 reflects two consecutive years of losses in foundation assets. Although assets began to rise again by the end of 2003, they remained below the peak reached in 2000. Moreover, there is usually a lag between changes in assets and changes in giving. Foundations—particularly those with large endowments—typically base their giving on the value of their assets averaged over two or more years.

The drop in giving in 2003 would have been greater were it not for at least three mitigating factors. First was the continued (albeit slower) growth in the number of foundations, which increased more than 1,500 in the latest year to over 66,000. Support from foundations reporting for the first time added $515.5 million to overall giving in 2003. Without the contribution of these new foundations, overall giving would have declined 2.1 percent, instead of 0.4 percent.

A second mitigating factor was the high level of new gifts into foundations, which increased 12.2 percent to $24.9 billion. A third factor in sustaining growth in giving was the continued commitment by many foundations to maintaining stable levels of giving—or at least curtail reductions—particularly in light of government cutbacks in social spending.

2003 Giving by Foundation Type. The slight decrease in giving in 2003 was concentrated among independent foundations, which comprise the vast majority of foundations. Giving by independent foundations—which includes family foundations and most of the "new health foundations"—decreased nearly 3 percent to $22.6 billion in the latest year. By comparison, giving by community and corporate foundations remained nearly unchanged (up 0.2 percent), while operating foundations raised their giving 46 percent.

Differences in the pattern of giving among foundation types partly reflect variations in their asset performance and gifts received. Compared to community and corporate foundations, independent foundations experienced a steeper decline in assets since 2001. Moreover, while new gifts into independent foundations rose 13.6 percent in 2003, this followed a sharp drop of 32.1 percent in the prior year. Community and corporate foundations experienced more moderate gains in gifts received in 2003—9.5 percent and 7.7 percent, respectively. But during the prior year, gifts into community foundations declined by only 0.3 percent, while gifts into corporate foundations slipped 1.3 percent.

TABLE 2.   Foundation Fiscal Data, 2002 to 2003

	2002	2003	% Change
No. of Foundations	64,843	66,398	2.4
Total Giving[1]	$ 30,431,799	$ 30,308,835	(0.4)
Total Assets	$435,190,471	$476,713,115	9.5
Gifts Received	$ 22,163,384	$ 24,858,326	12.2

Source: The Foundation Center, *Foundation Yearbook,* 2005. Dollars in thousands. Percent change represents current dollars.
[1]Includes grants, scholarships, and employee matching gifts; excludes set-asides, loans, PRIs, and program expenses.

FIGURE 8.   Effect of Inflation on Foundation Giving, 1975 to 2003

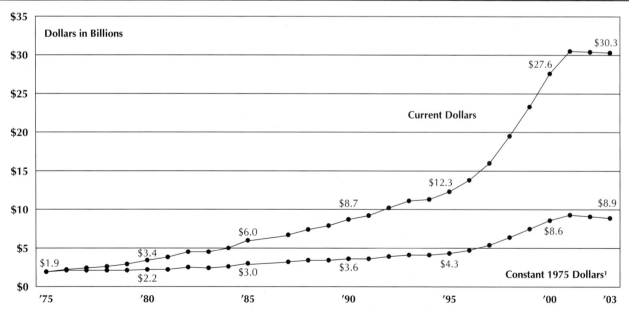

Source: The Foundation Center, *Foundation Yearbook,* 2005.
[1]Constant 1975 dollars based on annual average Consumer Price Index, all urban consumers, as reported by the U.S. Department of Labor, Bureau of Labor Statistics, as of February 2005.

Another factor contributing to the greater resiliency of giving by corporate and community foundations relative to independent foundations in 2003 is their higher growth in number. In the latest year, the number of independent foundations increased 2 percent, compared to a rise of nearly 8 percent for corporate foundations and 6 percent for community foundations. Excluding foundations reporting for the first time, 2003 giving by corporate and community foundations would have declined 0.7 percent and 0.5 percent, respectively.

Patterns of giving varied considerably for larger versus smaller funders. Among the different types of foundations, giving by the largest funders fared better than giving overall, although the changes in giving were highly uneven within the top groups. (For more detailed comparisons of giving by foundation type, see Chapter 4.)

2003 Giving by Region. Led by the American Contemporary Art Foundation (NY), Merck Patient Assistance Program (NJ), and Janssen Ortho Patience Assistance Foundation (NJ), grantmakers in the Northeast reported the fastest growth in giving. Overall, foundations based in the Northeast increased giving by 3.1 percent in 2003, following a 0.9 percent decline in the prior year. By comparison, giving by grantmakers in the West rose 1.3 percent, while giving decreased for foundations located in the South (down 2.4 percent) and the Midwest (down 5.4 percent). The Northeast also continued to account for the largest share of overall foundation giving (34.9 percent), followed by the South (22 percent). The West placed third (21.9 percent) and surpassed—for the first time on record—the Midwest (21.2 percent). (For more detailed comparisons of giving by region, see Chapter 3.)

Foundations' Share of Private Philanthropy, 2003

Sources of Private Contributions

- **Individuals provided over three-quarters of private contributions**

- **Excluding giving for religion, foundations accounted for roughly one-sixth of private contributions**

Given the large number of foundations, and the attention paid to a few highly visible grantmaker programs, grantseekers and journalists often overestimate the role that foundations play in the nonprofit sector. In fact, foundation giving represents a modest portion of all private contributions, which in turn account for a relatively small percentage of the overall income of America's nonprofits.

According to figures published in *Giving USA*, private contributions from all sources totaled $236.7 billion in 2003 (Table P1), representing an estimated 2.2 percent of the Gross Domestic Product.[1] The largest portion of these contributions—$198.7 billion or 84 percent—came from individual donors either through gifts or bequests (Figure P1). Independent, community, and grantmaking operating foundations were responsible for another 11.3 percent of total private giving, while the remaining 4.7 percent came from corporations and corporate foundations. (If cor-

porate foundation grant dollars are added to those of other foundations, the share rises to 12.8 percent for foundations.)

Individuals thus provide the vast majority of private donations, over five times that provided by foundations and businesses combined. Yet these proportions are somewhat misleading. It bears noting that well over one-third of giving from individuals is for the benefit of religious congregations, primarily, although far from exclusively, for sacramental purposes. Consequently, if religion is excluded from the private giving denominator, foundations' and corporations' share of support for the nonprofit public-benefit service sector increases significantly.[2]

For example, according to *Giving USA*, over one-third of private contributions in 2003 went to religion, most of it in gifts from individuals to their congregations. If religion is excluded, then independent and community foundations' share of the remaining contributions jumps to over 17 percent. While still relatively modest, this proportion nevertheless presents a far more accurate measure of all U.S. foundation support as a percent of overall private funding in the fields in which they are most active (e.g., education, human services, health, and the arts).

1. Giving USA Foundation, *Giving USA 2005: The Annual Report on Philanthropy for the Year 2004*, edited by M. Brown, Indianapolis, IN: Giving USA Foundation, 2005.

2. L. Salamon, *America's Nonprofit Sector: A Primer*, New York: Foundation Center, 1999.

Foundation giving as a proportion of the whole philanthropic pie grew rapidly during the 1960s and reached a peak in 1970 (9 percent). Throughout the 1970s, restrictive government regulations, inflation, and a shrinking asset base effectively reduced the role of foundations. By 1979, foundation's share of giving had dropped to 5.2 percent. With the easing of some government restrictions in the 1980s, a soaring stock market, and thousands of new foundations helping to raise asset values, foundation giving as a share of private contributions inched back up and continued to rise in the 1990s. In fact, between 1975 and 2003, inflation-adjusted growth in giving by independent and community foundations far outpaced increases reported for other types of private giving (Figure P2). However, independent and community foundations' share of all private philanthropic giving declined slightly in 2003. (Figure P3). Still, their current share of 11.3 percent represented the third highest share on record.

Corporate contributions, including corporate foundation giving, grew quickly between 1975 and 1986 and then experienced several years of decline. While support offered through corporate giving programs and corporate foundations grew in the 1990s, this growth often did not match increases reported by non-corporate foundations and individuals. Still, between 1975 and 2003, growth in corporate giving exceeded the overall increases reported for individuals and bequests during this period.

As a share of all private support, corporate philanthropy surpassed non-corporate foundations in 1984 (6 percent versus 5.8 percent). However, the 1987 stock market crash, corporate downsizing, deep losses in some industries during the recession of the early 1990s, numerous mergers, and continued challenges to corporate giving (often deemed by stockholders as too controversial or outside of the company's interests) pointed to a reduced role for business in American philanthropy. In addition, corporations are channeling an increasing share of their charitable support through corporate sponsorships and other forms of corporate marketing, which are not reflected in figures for charitable giving. (However, charitable giving does include in-kind giving, as well as cash support.) Since 2000, corporate contributions as a share of total private giving has hovered around 5 percent (Figure P3).

Sources of Income by Sub-sector

- **Private contributions accounted for less than one-fifth of nonprofits' income**

- **Private support most important for arts and civic affairs organizations**

Private giving from all sources—individuals, foundations, and corporations—accounts for only a fraction of overall nonprofit income. In 1998, nonprofit income (including church revenues), were estimated at $692.8 billion.[3] Of that total, one-eighth (12.5 percent) came from private contributions, compared to nearly four-fifths (79.5 percent) from earned income, including fees and other charges for services, income from endowments, and other receipts. The final one-twelfth (8 percent) came from federal, state, and local governments.

The proportion of annual income—both operating and capital—derived from these major sources

---

3. See Table 5.5 in M. Weitzman *et al.*, *The New Nonprofit Almanac and Desk Reference: The Essential Facts and Figures for Managers, Researchers, and Volunteers*, Washington, DC: Independent Sector, 2002.

FIGURE P1.　Distribution of Private Philanthropic Giving, 2003

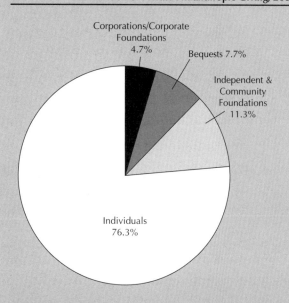

Corporations/Corporate Foundations 4.7%

Bequests 7.7%

Independent & Community Foundations 11.3%

Individuals 76.3%

Source: Giving USA Foundation, *Giving USA 2005*, ed. by M. Brown, Indianapolis, IN: Giving USA Foundation, 2005. Corporate data include corporate foundation giving.

TABLE P1.　Distribution of Private Philanthropic Giving, 2003

Source	Amount	%
Individuals	$180.6	76.3
Independent and Community Foundations	26.8	11.3
Bequests	18.1	7.7
Corporations/Corporate Foundations	11.2	4.7
Total	**$236.7**	**100.0**

Source: Giving USA Foundation, *Giving USA 2005*, ed. by M. Brown, Indianapolis, IN: Giving USA Foundation, 2005. Due to rounding, figures may not add up. Dollars in billions.
[1] Of total corporate giving for 2003, $3.5 billion (31.3%) was paid through corporate foundations (see Table 1).

of revenue differed widely by sector. For example, private donations in 1997 (the latest year for which comparable figures were available) accounted for 4 percent of health agencies' income, more than 13 percent for educational organizations, 20 percent for social service agencies, 36 percent for civic affairs organizations, and 44 percent for arts organizations. In the arts sub-sector, however, the high proportion of private contributions as a percent of income included gifts of expensive art objects donated at current market value to museums and other large capital gifts. If those gifts were excluded, private donations would represent a smaller share of income. Still, foundation giving, as one source of private support, is likely to be far more important to the average arts organization than to the average health organization.

FIGURE P2.   Change in Private Philanthropic Giving by Source, 1975 to 2003

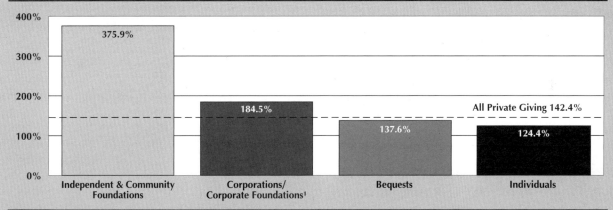

Source: Giving USA Foundation, *Giving USA 2005*, ed. by M. Brown, Indianapolis, IN: Giving USA Foundation, 2005. Figures based on constant dollars.
[1]Figures include direct corporate giving as well as giving by corporate foundations.

FIGURE P3.   Giving by Independent and Community Foundations and Corporations/Corporate Foundations as a Share of All Private Giving, 1975 to 2003

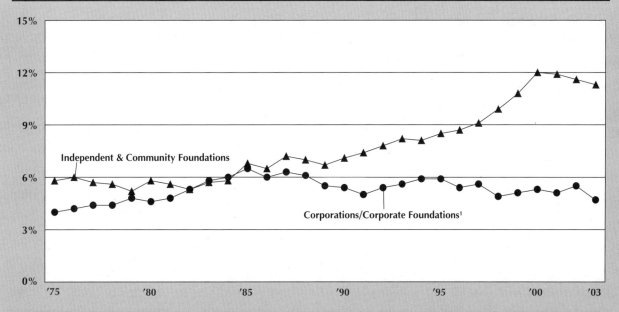

Source: Giving USA Foundation, *Giving USA 2005*, ed. by M. Brown, Indianapolis, IN: Giving USA Foundation, 2005. Over four out of five private philanthropic dollars are provided each year by living donors and through bequests.
[1]Figures include direct corporate giving as well as giving by corporate foundations.

2003 Assets

- Foundation assets rebounded to $476.7 billion, up nearly 10 percent

- Asset levels remained below 2000 peak

A strong recovery in the U.S. stock market, augmented by an increase in new gifts into foundations, resulted in greater foundation resources in 2003. Overall, foundation assets rose 9.5 percent to $476.7 billion in the latest year (Table 2), ending two consecutive years of losses. Foundation assets, however, remained below the peak of $486.1 billion reached in 2000. Still, between 1995 and 2003, assets grew by $250 billion, representing an inflation-adjusted increase of 74.1 percent (Figure 9).

Among the major factors affecting the value of foundation assets—especially of the largest U.S. foundations—was the performance of the stock market. Changes in the value of foundation endowments typically follow the direction of U.S. investment markets, although they are generally less extreme. For example, at the close of 2003, all of the major U.S. stock market indices registered double-digit rates of growth, with the technology-heavy NASDAQ index posting the largest gain (up 50 percent), followed by the Wilshire 5000 stock index (up 29.4 percent), the New York Stock Exchange composite index (up 29.3 percent), the Standard and Poor's 500 stock index (up 26.4 percent), and the Dow Jones industrial average

(up 25.3 percent). Not surprisingly, foundation endowments also rose in 2003, although at a much more modest rate (9.5 percent) than the top-performing stock market benchmarks.

Part of the difference in returns is attributable to the fact that only a portion of foundation assets are invested in the U.S. stock market.[2] Large endowed foundations tend to hold more of their assets in equities than smaller foundations. Not surprisingly, during years of strong market growth, the largest endowed foundations generally outperform mid-size and smaller foundations in terms of asset gains. In the reinvigorated 2003 market, total assets of the very largest foundations (those with assets of $250 million or more) increased nearly 13 percent, while the smaller foundations (those with assets of $1 million or less) experienced 1.4 percent growth in their total assets.

2003 Assets by Foundation Type. The four types of foundations covered in this study all showed asset gains in 2003. Community foundations reported the largest increase—14.7 percent, to $34.2 billion. Large and medium-sized community foundations were responsible for the lion's share of this growth. Assets of independent foundations rose 9.6 percent to $399.1 billion. Contributing to this increase was a 13.6 percent gain in new gifts into independent foundations. Corporate foundation assets grew 7.1 percent. This lower-than-average growth reflected in part the high level of giving (payout) relative to gifts into their foundations

FIGURE 9. Effect of Inflation on Foundation Assets, 1975 to 2003

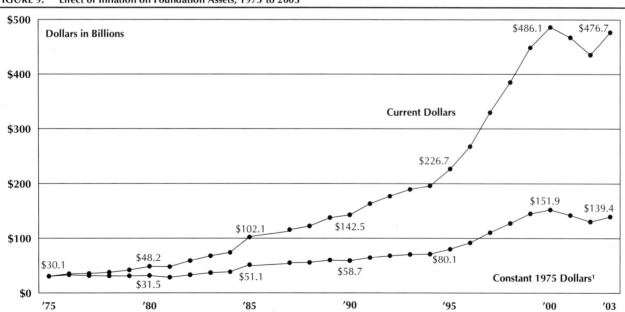

Source: The Foundation Center, *Foundation Yearbook,* 2005.
[1]Constant 1975 dollars based on annual average Consumer Price Index, all urban consumers, as reported by the U.S. Department of Labor, Bureau of Labor Statistics, as of February 2005.

(pay-in). In 2003, corporate foundation giving exceeded gifts received by $231.5 million. Helping to offset this high payout and lifting corporate foundation assets were the 187 corporate foundations new to the pool of active grantmakers in 2003. These new foundations held $187.1 million in assets. Excluding these new foundations, corporate assets would have increased 5.8 percent, instead of 7.1 percent. (For more detailed comparisons of assets by foundation type, see Chapter 4.)

2003 Assets by Region. Foundations in the West experienced the strongest growth in assets among the four major regions. Overall, the assets of foundations in the West grew 14.3 percent between 2002 and 2003, compared with corresponding increases of 8.7 percent in the South, 7.9 percent in the Midwest, and 7.3 percent in the Northeast. Nonetheless, the Northeast still accounted for the largest share of overall foundation assets (30 percent), followed by the West (27.1 percent), Midwest (21.7 percent), and South (21.2 percent). (For more detailed comparisons of assets and gifts by region, see Chapter 3.)

2003 Gifts Received

- **Gifts and bequests from donors to foundations rose 12 percent**

- **New gifts totaled $24.9 billion**

- **Independent foundations reported largest increase in new gifts**

Variations in the growth patterns of grantmakers often result from new gifts to endowments or from differences in the funding policies of endowed and non-endowed (sometimes called "pass-through") foundations. Grants of endowed foundations are generally paid from investment income, while principal is left to grow. In contrast, many corporate foundations and family foundations with living donors maintain only a small asset base, instead funding grants from annual gifts.

Gifts and bequests from donors to their foundations rose 12.2 percent in 2003, from $22.2 billion to $24.9 billion (Table 2). This increase followed a decline of nearly 23 percent in the prior year. New gifts in 2003 were boosted by the $3.7 billion gift to the Gordon and Betty Moore Foundation from its founders. In addition, the $24.9 billion total also included $1.2 billion in gifts from donors to new foundations. Still, overall gifts into foundations remained below the peak of $32.1 billion, reached in 1999.

Unlike for giving and assets, growth of new gifts is unpredictable. Typically, the largest gifts are triggered by one-time events, such as a donor's death. Many of the largest gifts to foundations in recent years resulted from bequests, and the frequency of such large gifts has increased over the past several years. This suggests that foundations may be experiencing the much-publicized transfer of wealth that is expected as the World War II generation passes on. Evidence supporting this development includes a fivefold increase in the annual amount of gifts and bequests since 1990.

Number of Large Gifts. Of the nation's 66,398 active grantmaking foundations, 27,918 (42 percent) received gifts from donors in 2003, but only a relatively few received substantial gifts. In the latest year, 689 foundations received at least $5 million in gifts and bequests (Table 3), a decrease of 3.4 percent from 2002. Of these foundations, 371 received gifts and bequests of $10 million or more, a 1.1 percent decline from 2002. The decline in the number of foundations receiving very large gifts was greatest for operating foundations (down 7.7 percent), followed by independent foundations (down 6.5 percent). In contrast, the number of community foundations receiving gifts of $10 million or more rose 14.7 percent, and corporate foundations experienced a 3.1 percent increase.

Total Value of Large Gifts. Although fewer foundations overall received large gifts, the total value of these gifts rose substantially. The value of gifts of $5 million or more grew 18 percent in 2003, from $15.1 billion to $17.9 billion. The increase was even greater for the total amount of gifts of $10 million—up 22.2 percent, from $12.9 billion to $15.7 billion. Lists of the top foundations by gifts received are available in the FC Stats section of the Foundation Center's Web site (see www.fdncenter/fc_stats).

Among the four major foundation types, independent foundations posted the largest increase in total gifts received. Overall, gifts to independent foundations rose 13.6 percent in 2003, from $14 billion to $15.8 billion, following a 32.1 percent decline in the

TABLE 3. Number of Foundations Receiving Total Gifts of $5 Million or More, 2003

	Size of Total Gifts Received			
	$5 million or more		$10 million or more	
Foundation Type	No. of Foundations	Amount	No. of Foundations	Amount
Independent	400	$10,748,957	202	$ 9,425,665
Corporate	126	2,511,035	67	2,104,712
Community	119	2,854,580	78	2,560,759
Operating	44	1,764,334	24	1,623,654
Total	689	$17,878,906	371	$15,714,790

Source: The Foundation Center, *Foundation Yearbook,* 2005.
   Dollars in thousands.

prior year. Gifts into community foundations increased 9.5 percent to $3.5 billion. Similarly, gifts from companies into their foundations rose 7.7 percent to $3.2 billion in 2003. (For more detailed comparisons of gifts received by foundation type, see Chapter 4.)

Gifts Received by Region. Most of the increase in gifts received in the latest year was concentrated among foundations in the West. Overall, gifts into foundations in the West jumped 55.5 percent to $7.4 billion in 2003, largely due to the exceptional $3.7 billion gift into the California-based Gordon and Betty Moore Foundation. Foundations in the South also reported a strong increase of 23.2 percent in gifts received to $6.3 billion. In contrast, Midwestern foundations posted an 11.4 percent decline to $4.4 billion, while Northeastern foundations registered a 7.8 percent decrease to $6.8 billion. (For more detailed comparisons of gifts received by region, see Chapter 3.)

Top 25 Foundations by Giving and Assets

A comparison of 2002 and 2003 financial data for the 25 largest foundations by giving (Table 4) showed a 10 percent overall increase in giving, compared to a 0.4 percent dip for foundations overall. Sixteen of the top 25 foundations reported growth in giving, while the remainder posted declines.

Many of the largest foundations saw their assets recover in 2003. A comparison of the asset values of the 25 largest foundations by assets showed a nearly 17 percent rise in their endowments between 2002 and 2003 (Table 5)—well above the 9.5 percent increase in assets reported for grantmaking foundations overall. Assets grew by at least ten percent for 18 of the top 25 foundations. Six funders registered single digit increases, while only one foundation reported a decrease.

Giving

- Giving by the top 25 funders increased 10 percent

- Rise in giving by top funders contrasted with decrease in overall giving

- Median change in giving for top 25 was a 9.3 percent increase

From 2002 to 2003, giving by the nation's top 25 grantmakers by size of giving rose 10 percent, compared to a 0.4 percent decrease reported for all 66,398 grantmaking foundations. Total giving by the largest funders increased from $6 billion to $6.6 billion. These 25 funders accounted for over one-fifth (21.7 percent) of the $30.3 billion in overall 2003 giving. By comparison,

the top 25 funders by size of giving in 2002 accounted for a slightly smaller share (20.3 percent) of the $30.4 billion in giving reported that year.

Although overall giving increased among the top 25 foundations in 2003, there was considerable variation within the group. Sixteen foundations posted upturns in giving, while nine reported declines. Three foundations experienced exceptional growth in their giving, led by the American Contemporary Art Foundation (NY), with giving up more than 800-fold to $192.7 million; Lilly Cares Foundation (IN), with a more than 12-fold jump in giving to $113.3 million; and F.W. Olin Foundation (FL), with giving more than doubling to

TABLE 4.   Comparison of the 25 Largest Foundations by Total Giving, 2002 to 2003

Foundation	Total Giving 2002[1]	Total Giving 2003[1]	% Change	Rank 2002
1. Bill & Melinda Gates Foundation	$1,158,292	**$1,182,827**	2.1	1
2. Lilly Endowment	557,098	**462,337**	(17.0)	2
3. Ford Foundation	509,700	**431,643**	(15.3)	3
4. Bristol-Myers Squibb Patient Assistance Foundation[2]	297,135	**401,840**	35.2	6
5. Robert Wood Johnson Foundation	360,347	**390,600**	8.4	4
6. Merck Patient Assistance Program[2]	244,000	**380,860**	56.1	N/A
7. David and Lucile Packard Foundation	350,048	**277,892**	(20.6)	5
8. Pew Charitable Trusts[3]	238,535	**265,438**	11.3	7
9. William and Flora Hewlett Foundation	168,214	**258,048**	53.4	13
10. Janssen Ortho Patient Assistance Foundation[2]	155,304	**214,136**	37.9	15
11. Annenberg Foundation	195,071	**213,210**	9.3	11
12. John D. and Catherine T. MacArthur Foundation	195,573	**195,115**	(0.2)	10
13. W. K. Kellogg Foundation	176,303	**195,007**	10.6	12
14. American Contemporary Art Foundation[2]	229	**192,678**	84,038.9	N/A
15. Starr Foundation	209,301	**188,856**	(9.8)	9
16. Andrew W. Mellon Foundation	222,662	**179,159**	(19.5)	8
17. Annie E. Casey Foundation	159,310	**172,797**	8.5	14
18. California Endowment	153,441	**152,021**	(0.9)	16
19. Rockefeller Foundation	149,160	**130,081**	(12.8)	17
20. Wal-Mart Foundation	101,000	**119,801**	18.6	23
21. New York Community Trust	126,485	**117,992**	(6.7)	19
22. Kresge Foundation	98,974	**116,695**	17.9	25
23. Lilly Cares Foundation	9,306	**113,311**	1,117.6	426
24. F. W. Olin Foundation	54,911	**110,621**	101.5	57
25. Walton Family Foundation	80,453	**106,903**	32.9	36
Total	**$5,970,852**	**$6,569,868**	**10.0**	

Source: The Foundation Center, *Foundation Yearbook,* 2005. Aggregate foundation fiscal information in tables and figures in this report is based on data provided to the Center as of January 2005. Fiscal data on individual foundations included in this table may be more current. Dollars in thousands.
[1] Includes grants, scholarships, and employee matching gifts; excludes set-asides, loans, PRIs and program expenses.
[2] For some operating foundations, program expenses are included.
[3] As of January 1, 2004, the Pew Charitable Trusts changed its status to a public charity.
N/A = Not available.

$110.6 million. Nine foundations posted double-digit growth in giving. The biggest increases among this group were reported by the Merck Patient Assistance Program (NJ), with giving up 56.1 percent to $380.9 million, followed by the William and Flora Hewlett Foundation (CA), with giving up 53.4 percent to $258 million. In contrast, five foundations registered double-digit declines in giving, with the David and Lucile Packard Foundation showing the steepest drop (down 20.6 percent to $277.9 million), while four posted decreases of under 10 percent.

A more equitable measure of change focuses on the median giving amounts. For the top funders, the median change in 2003 was a 9.3 percent increase, compared to a 1 percent uptick in 2002 and a 2.4 percent rise in 2001.

TABLE 5. Comparison of the 25 Largest Foundations by Assets, 2002 to 2003

Foundation	Assets 2002	Assets 2003	% Change	Rank 2002
1. Bill & Melinda Gates Foundation	$ 24,081,369	$ 26,810,518	11.3	1
2. Lilly Endowment	10,054,032	10,849,402	7.9	2
3. Ford Foundation	9,345,030	10,015,613	7.2	3
4. J. Paul Getty Trust	8,623,796	9,100,188	5.5	4
5. Robert Wood Johnson Foundation	8,012,367	7,933,761	(1.0)	5
6. W. K. Kellogg Foundation	5,729,303	6,801,844	18.7	6
7. William and Flora Hewlett Foundation	5,010,197	6,190,363	23.6	7
8. David and Lucile Packard Foundation	4,793,893	5,982,468	24.8	8
9. Gordon and Betty Moore Foundation	93,323	4,846,115	5,092.8	547
10. Andrew W. Mellon Foundation	3,600,620	4,719,646	31.1	11
11. John D. and Catherine T. MacArthur Foundation	3,836,622	4,530,411	18.1	9
12. Pew Charitable Trusts[1]	3,753,638	4,118,768	9.7	10
13. Starr Foundation	3,322,103	3,577,379	7.7	12
14. California Endowment	2,762,621	3,572,425	29.3	13
15. Annie E. Casey Foundation	2,709,413	3,106,521	14.7	14
16. Rockefeller Foundation	2,679,064	3,026,669	13.0	15
17. Annenberg Foundation	2,331,680	2,695,762	15.6	16
18. Kresge Foundation	2,164,478	2,504,554	15.7	18
19. Robert W. Woodruff Foundation	2,210,194	2,464,363	11.5	17
20. Charles Stewart Mott Foundation	2,011,396	2,376,108	18.1	20
21. Duke Endowment	2,084,822	2,307,707	10.7	19
22. Harry and Jeanette Weinberg Foundation	1,792,786	1,986,758	10.8	22
23. McKnight Foundation	1,549,715	1,914,001	23.5	28
24. John S. and James L. Knight Foundation	1,718,236	1,845,869	7.4	24
25. Carnegie Corporation of New York	1,627,734	1,824,315	12.1	25
Total	**$115,898,432**	**$135,101,530**	**16.6**	

Source: The Foundation Center, *Foundation Yearbook*, 2005. Aggregate foundation fiscal information in tables and figures in this report is based on data provided to the Center as of January 2005. Fiscal data on individual foundations included in this table may be more current. Dollars in thousands.
[1] As of January 1, 2004, the Pew Charitable Trusts changed its status to a public charity.

Variations in giving patterns among the leading funders can be attributed to differences in their investment performance over the most recent fiscal periods; dramatic influxes of assets through gifts and bequests from donors; or variations in fiscal reporting periods. They may also reflect differences in long-term grant distribution strategies or one-time distributions that skew funding levels for a single year. For example, the extraordinary jump in giving in 2003 by the American Contemporary Art Foundation was due to a $189.8 million grant in the form of art work awarded to the Whitney Museum of American Art. The huge increase in the Lilly Cares Foundation's giving consisted of pharmaceutical products supplied by its corporate sponsor, Eli Lilly and Company. In the case of the F.W. Olin Foundation, the doubling of its giving was attributable to a $106.6 million award to the Franklin W. Olin College of Engineering (MA). The foundation, which was established in 1938, plans to spend down its assets and terminate by the end of 2005.

For the fourth consecutive year, the Gates Foundation ranked first by total giving. The Lilly Endowment and the Ford Foundation placed second and third, respectively—unchanged from 2002. Rankings of the remaining top ten foundations shifted in the latest year. Among those rising in rank were the Bristol-Myers Squibb Patient Assistance Foundation, Merck Patient Assistance Program, William and Flora Hewlett Foundation, and Janssen Ortho Patient Assistance Foundation. Three of these foundations were new to the top ten in the latest year. Merck was unranked in 2002, but placed sixth in the latest year. Hewlett rose from thirteenth to ninth place, while Janssen climbed from fifteenth to tenth place. Displaced from the top ten were the MacArthur, Starr, and Mellon foundations—but all three remained among the top 25 foundations by total giving.

Assets

- **Assets of the 25 largest foundations rose nearly 17 percent**

- **Median asset increase was 13 percent**

- **Among top 25, Moore Foundation showed the biggest jump in assets**

From 2002 to 2003, the combined assets of the 25 largest endowed foundations increased 16.6 percent, surpassing the 9.5 percent rise in assets reported for foundations overall in the latest year. By dollar value, their endowments grew from $115.9 billion to $135.1 billion. Six foundations reported gains of more than $1 billion, led by the Gordon and Betty Moore Foundation, whose assets ballooned by nearly $4.8 billion, followed by Gates Foundation, which saw its endowment grow by $2.7 billion. Among the top 25 foundations, only one posted a decline in assets—the Robert Wood Johnson Foundation, with assets slipping 1 percent to $7.9 billion.

The 16.6 percent rise in assets among the top 25 foundations fell below the average gains in the major stock market indexes in 2003, which ranged from a low of 25.3 percent for the Dow Jones industrial average to a high of 50 percent for the NASDAQ. Because many of the largest foundations maintain diversified investment portfolios, changes in their assets tend to be less volatile than the major market benchmarks. Moreover, new gifts into foundations, as well as payout levels, also affect asset performance.

Among grantmaking foundations, however, the largest endowed foundations tend to outperform smaller foundations in terms of asset growth during periods of strong market growth. Nineteen of the 25 largest foundations surpassed the 9.5 percent average increase in assets for all grantmaking foundations in 2003. Aside from the Moore Foundation, the largest percentage gains were reported by the Mellon Foundation (up 31.1 percent to $4.7 billion), followed by the California Endowment (up 29.3 percent to $3.6 billion). Nonetheless, the assets of these two foundations—as well as those of many other top funders—remained below the peak levels recorded in 2000.

Two of the top ten foundations by assets improved their standing in 2003. Due to the exceptional $3.7 billion infusion from their founders, the Gordon and Betty Moore Foundation's ranking jumped from 547th to ninth. The A.W. Mellon Foundation also increased in rank from eleventh to tenth place. Displaced from the top ten in the latest year were the MacArthur Foundation, which slipped from ninth to eleventh

place, and the Pew Charitable Trusts, which fell from tenth to twelfth place.

Growth of the Top Ten Foundations, 1993 to 2003

Most of the nation's largest philanthropies were created and received their principal endowments long ago. Nevertheless, new formation, new bequests, and a boom in investment markets in the late 1990s brought about changes in the group of top foundations.

A comparison of the ten largest funders of 1993 by asset size (Table 6) with the largest foundations of 2003 (Table 5) showed that four foundations have been added and four displaced in recent years. The additions include the Bill & Melinda Gates Foundation (ranked first in 2003), William and Flora Hewlett Foundation (ranked seventh), David and Lucile Packard Foundation (ranked eighth), and Gordon and Betty Moore Foundation (ranked ninth). The Gates Foundation received the largest gift into a foundation in 1998, 1999, and 2000 and the second largest gift in 2001; the Hewlett Foundation received the largest gift in 2001 through a bequest from its founder; the Packard Foundation experienced spectacular growth in the late 1990s based on both a bequest and investment performance through 1999; and the Moore Foundation received the largest gift into a foundation in 2003. Foundations displaced from the top ten included the John D. and Catherine T. MacArthur Foundation (ranked eleventh), the Pew Charitable Trusts (ranked twelfth), the Rockefeller Foundation (ranked sixteenth), and the Robert W. Woodruff Foundation (ranked nineteenth). Among the six remaining

TABLE 6.   Ten Largest Foundations by Assets in 1993: Change in Assets and Total Giving, 1993 to 2003

Foundation Name	Assets			Rank by Assets 2003	Total Giving[1]		
	1993	2003	% Change		1993	2003	% Change
1. Ford Foundation	$6,955,937	$10,015,613	44.0	3	$281,859	$431,643	53.1
2. J. Paul Getty Trust[2]	6,184,253	9,100,188	47.2	4	9,924	19,308	94.6
3. W. K. Kellogg Foundation	5,046,557	6,801,844	34.8	6	227,448	195,007	(14.3)
4. Pew Charitable Trusts[3]	3,512,045	4,118,768	17.3	12	227,448	265,438	16.7
5. Robert Wood Johnson Foundation	3,461,378	7,933,761	129.2	5	143,386	390,600	172.4
6. John D. and Catherine T. MacArthur Foundation	3,098,880	4,530,411	46.2	11	105,834	195,115	84.4
7. Lilly Endowment	2,520,098	10,849,402	330.5	2	132,486	462,337	249.0
8. Rockefeller Foundation	2,364,553	3,026,669	28.0	16	93,261	130,081	39.5
9. Andrew W. Mellon Foundation	2,330,432	4,719,646	102.5	10	93,990	179,159	90.6
10. Robert W. Woodruff Foundation	1,594,387	2,464,363	54.6	19	52,741	102,637	94.6

Source: The Foundation Center, *Foundation Yearbook*, 2005. Dollars in thousands.

[1]Includes grants, scholarships, and employee matching gifts; excludes set-asides, loans, PRIs, and program expenses. For some operating foundations, program expenses are included.

[2]The J. Paul Getty Trust, an operating foundation, is not subject to the same charitable distribution requirement as independent foundations.

[3]As of January 1, 2004, the Pew Charitable Trusts changed its status to a public charity.

foundations, Ford consistently ranked first until 1997, when the Lilly Endowment displaced it. In 1999, Lilly was displaced by Gates. Rankings of the other funders have fluctuated due to differences in asset growth patterns.

Among the top ten foundations by asset size in 1993, the Lilly Endowment showed the largest percentage increase in assets over the decade. Lilly's assets more than quadrupled from $2.5 billion to $10.8 billion in 2003; asset rank rose from seventh to second; and giving more than tripled, from $132.5 million to $462.3 million. The second largest percentage increase in assets over the decade was posted by the Robert Wood Johnson Foundation, and its giving nearly tripled, from $143.4 million to $390.6 million. The other foundations experienced growth in their assets ranging from 17 percent to over 102 percent. For seven out of the ten foundations (the exceptions including the Kellogg, Pew, and Mellon foundations), the gains in giving far outstripped their asset growth.

Large and Small Foundations

• **About one-third of foundations controlled over 97 percent of assets**

• **Only 380 foundations awarded grants totaling at least $10 million**

• **Larger foundations were more likely to fund nationally and internationally**

Of the 66,398 active grantmaking foundations, 23,142 (34.9 percent) held assets of $1 million or more in 2003 (Table 7). This relatively small share of foundations accounted for over 97 percent of total foundation assets and 90 percent of giving. The number of large or very large foundations (those with assets of $50 million or over) totaled 1,115 in 2003. This group held 69 percent of assets (Figure 10) and was responsible for 54 percent of all giving. The smallest 43,000 foundations (those with assets of less than $1 million) held roughly 3 percent of assets but accounted for about 10 percent of giving.

A more revealing measure of foundation size, especially for non-endowed foundations, is total contributions. As shown in Table 8, only 380 foundations (just over half of 1 percent) reported grants of $10 million or over, yet together they accounted for more than half (51.3 percent) of giving. Close to 21,000 foundations (31.2 percent) awarded $100,000 or more in grants; their combined support represented 96 percent of giving.

Figures 11 and 12 show the share of giving and assets held in 2003 by a fixed number of foundations—the 1,000 largest, 100 largest, and 50 largest. The graphs reveal that assets were concentrated among even fewer large foundations than were total grants. For example, the 1,000 largest foundations by asset size controlled 67 percent of asset dollars, while the

FIGURE 10. Distribution of Foundations and Foundation Assets by Asset Range, 2003

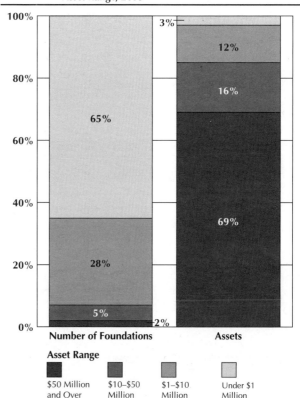

Source: The Foundation Center, *Foundation Yearbook*, 2005.

TABLE 7. Analysis of Foundations by Asset Size, 2003

Asset Range	Number of Foundations	%	Assets	%	Total Giving[1]	%
$1 billion+	46	0.1	$163,377,516	34.3	$6,507,666	21.5
$250 million–$1 billion	161	0.2	74,357,687	15.6	4,013,828	13.2
$50 million–$250 million	908	1.4	92,458,500	19.4	5,974,240	19.7
$10 million–$50 million	3,617	5.4	77,346,419	16.2	5,307,777	17.5
$1 million–$10 million	18,410	27.7	57,342,778	12.0	5,499,923	18.1
Under $1 million	43,256	65.1	11,830,215	2.5	3,005,401	9.9
Total	**66,398**	**100.0**	**$476,713,115**	**100.0**	**$30,308,835**	**100.0**

Source: The Foundation Center, *Foundation Yearbook*, 2005. Dollars in thousands. Due to rounding, figures may not add up.
[1]Includes grants, scholarships, and employee matching gifts; excludes set–asides, loans, PRIs, and program expenses.

1,000 largest by size of giving distributed 63 percent of all grant dollars. The 100 largest grantmakers by asset size controlled 42 percent of asset dollars, yet the 100 largest foundations ranked by total giving paid out 35 percent of grant dollars. Finally, the 50 largest foundations by assets held 35 percent of assets, while the 50 largest foundations by total giving distributed less than 30 percent of grant dollars.

Because they control a substantial portion of funds, the largest foundations merit close scrutiny. The grant programs of the 1,000 largest foundations by giving are tracked in the Foundation Center's *Foundation 1000*. Giving patterns of over 1,000 of the largest foundations are analyzed in *Foundation Giving Trends*, part of the *Foundations Today Series* of annual research reports. Finally, rankings of the 50 largest foundations by giving, assets, and gifts received are available in FC Stats, the Foundation Center's online statistical information service (see www.fdncenter/fc_stats).

There are significant differences between these few larger foundations and the vast majority of private foundations, most importantly:

Local vs. National Giving. The overwhelming majority of the nation's 66,398 active foundations limit their giving to their local community. Fewer than one out of fifteen grantmaking foundations (4,347) give on a national or international basis, and they are usually the largest independent and operating foundations, or corporate foundations whose sponsoring companies operate nationally or internationally. Although few in number, national foundations accounted for 36 percent of the assets held by all grantmaking foundations in 2003 and were responsible for 33 percent of giving in that year.

TABLE 8.   Analysis of Foundations by Total Giving Range, 2003

Total Giving Range	Number of Foundations	%	Total Giving[1]	%
$100 million+	32	0.0	$ 7.0	23.1
$25 million–$100 million	108	0.2	4.8	15.9
$10 million–$25 million	240	0.4	3.7	12.4
$1 million–$10 million	3,197	4.8	8.6	28.2
$100,000–$1 million	17,136	25.8	5.0	16.4
Under $100,000	45,685	68.8	1.2	4.0
Total	**66,398**	**100.0**	**$30.3**	**100.0**

Source: The Foundation Center, *Foundation Yearbook*, 2005. Dollars in billions. Due to rounding, figures may not add up.
[1]Includes grants, scholarships, and employee matching gifts; excludes set–asides, loans, PRIs, and program expenses.

FIGURE 11.   Largest Foundations by Total Giving, 2003

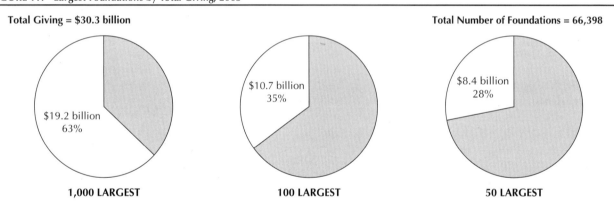

Source: The Foundation Center, *Foundation Yearbook*, 2005.

FIGURE 12.   Largest Foundations by Asset Size, 2003

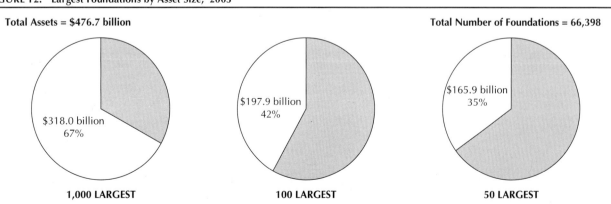

Source: The Foundation Center, *Foundation Yearbook*, 2005.

Broad Giving vs. Defined Programs. Many of the largest foundations organize their giving through announced programs that may be limited to a specific subject or field (e.g., the Robert Wood Johnson Foundation concentrates on the health field), or cover a wide range of interests. Small foundations that give locally tend to support a broad range of activities. Information about their programs is generally not as widely available to the public. Nevertheless, these funders remain an important source of grants for community service agencies, cultural organizations, schools, and other locally based nonprofit organizations.

Public Reporting. Nearly all of the 66,398 grantmaking foundations are required by law to file IRS Form 990-PF and must make that document available to the public. In addition, roughly 3,300 foundations issue statements of their program interests or guidelines for grant applications, and about 1,500—mainly the larger staffed foundations—state that they publish annual or biennial reports. Of the smaller grantmakers publishing reports, many are community trusts. Community trusts are not classified as private foundations by the IRS and are therefore not required to follow the same public reporting requirements as private foundations. Instead, most report their giving voluntarily. Finally, in addition to published reports, more than 2,600 private and public foundations currently maintain Web sites, which are accessible via the Foundation Center's Web site at www.fdncenter.org.

Staffing. Only the largest foundations are likely to have paid staff to review and investigate proposals, develop projects, and work with the public. From a survey mailed to nearly 20,000 larger foundations by the Foundation Center in 2004, only 3,353 respondents (roughly one-in-six) reported paid staff. Moreover, of those with assets lower than $5 million, only about one-in-thirteen reported paid staff. Few, if any, of the more than 46,000 small foundations not included in the Foundation Center's survey are able to employ paid staff. Instead, their programs are typically administered by lawyers, bank trustees, and family members on a part-time basis.

Measures of Foundation Support, 2003

- **2003 qualifying distributions totaled $33.9 billion**

- **Grants to individuals totaled almost $1.1 billion**

- **Loans and other charitable investments totaled $190.4 million**

- **Top 20,800 foundations awarded more than one million grants**

If asked how to measure the extent of foundations' charitable support, most people would suggest counting up all of the money that funders give to organizations in the form of grants. While this measure would accurately reflect the vast majority of foundations' charitable activities, it would by no means capture all of their philanthropic support.

Foundations employ a number of tools to promote their charitable mission. For example, beyond making grants to organizations, funders can make charitable loans and other forms of below-market-rate investments also known as program-related investments (PRIs) as well as grants and loans to individuals. These types of charitable support provide several measures of foundation philanthropy. Moreover, minimum levels of private foundation support have been set by the Federal government, providing another standard for measuring foundations' charitable activities.

The following sections detail several ways, beyond grantmaking to organizations, by which to measure foundation support:

Qualifying Distributions. Qualifying distributions represent the total amount that foundations disbursed for charitable activities in a given fiscal year that can be counted toward their Federal payout requirement (see below). Grants to organizations constitute the vast majority of qualifying distributions—an amount that also includes grants to individuals, charitable investments, set asides, most operating expenses, and other charitable expenditures, e.g., the costs for convening a conference on the U.S. health care system, offering office space to nonprofits, and providing technical assistance. In 2003, qualifying distributions totaled $33.9 billion (Table 1).

Payout. U.S. law requires that all private, non-operating foundations pay out a minimum of 5 percent of the value of their assets in the preceding year as charitable distributions.[1] While we cannot precisely measure aggregate payout, 2003 giving

---

1. The principal purpose of operating foundations is to conduct their own programs or provide a direct service. Therefore, they are not required to meet the same payout requirement as other private foundations.

represented 6.2 percent of independent foundation assets at the close of 2002, up from 5.9 percent in the previous year (see Figure 5 in Chapter 1). For all private (non-operating) foundations, which includes independent and corporate foundations, the ratio inched up from 6.5 percent in 2002 to 6.9 percent in the latest year. The ratio of independent and corporate foundations' qualifying distributions to their 2002 assets was 7.5 percent.[2]

The payout ratio varies by foundation type and size. In general, corporate foundations and smaller foundations that do not maintain endowments pay out at a higher rate than the large endowed foundations. In contrast, most large independent foundations, with income derived solely from their endowments, adhere to a 5 percent payout strategy, in compliance with IRS regulations.[3]

Grants to Individuals. Of the total $30.3 billion in foundation giving awarded in 2003, the Foundation Center identified $1.1 billion in scholarships and other grants to individuals, up 35.4 percent from $809.9 million in 2002. While the total amount has increased substantially since the early 1990s, grants paid directly to individuals represented a tiny share of overall foundation giving (3.6 percent). Still, in 2003 over 362,000 individuals, more than double the number in 2002, benefited from grants and scholarships made directly by foundations.[4]

Program-Related Investments (PRIs). Foundations provided $190.4 million in low-interest or interest-free loans and charitable investments known as PRIs for projects closely related to their grantmaking interests (Table 1). The use of this type of alternative financing decreased a substantial 23.7 percent in 2003, from $249.4 million in 2002. Most of this decline was attributable to the drop in the level of PRIs awarded by operating foundations, from $57.6 million in 2002 to $10.9 million in 2003. Some of the leading independent foundation PRI providers in the latest year included the David and Lucile Packard, Ford and John D. and Catherine T. MacArthur foundations. Notable corporate foundations included the Prudential and Metlife foundations. A number of community foundations, such as the Cleveland foundation, operate small loan programs for local nonprofits.[5] Finally, although the amount of PRI financing may fluctuate year to year, the number of active PRI providers has remained fairly stable.

Number of Grants. Foundation Center staff identified 1.2 million grants made by the roughly 20,800 foundations with assets of at least $1 million or grants of at least $100,000.[6] Independent foundations accounted for more than half (54.7 percent) of these grants, corporate foundations (31.4 percent) and community foundations (12.7 percent). Only 1.2 percent of grants were made by operating foundations.

2. These ratios are merely illustrative and do not equal actual payout, which is calculated individually for each foundation based on a five-year formula that allows for carrying forward distributions in excess of the 5 percent requirement.
3. A 2000 study, "Spending Policies and Investment Planning for Foundations," assessed the impact of foundation spending policies on the value of their portfolios. The report, prepared by DeMarche Associates for the Council on Foundations, concluded that foundations could not maintain payout rates of 6 percent or higher without undermining the purchasing power of their portfolios. A 2000 study prepared by Cambridge Associates for the Council of Michigan Foundations, "Sustainable Payout for Foundations," reached a similar conclusion.

4. Although the Foundation Center does not maintain comprehensive data on individual giving of all foundations, descriptive entries on more than 6,000 large and small grantmakers that award grants to individuals are currently provided in *Foundation Grants to Individuals Online* at www.gtionline.fdncenter.org.
5. The latest information on charitable investing was released in the Foundation Center's 2003 publication, *The PRI Directory: Charitable Loans and Other Program-Related Investments by Foundations,* 2nd Edition.
6. Unlike information on grant dollars awarded, data on the number of grants awarded may be incomplete for a few funders. Therefore, the total figure for number of grants awarded by these foundations is probably somewhat higher.

Endnotes

1. Giving represents payments, not necessarily all new authorizations. It includes grants to organizations, scholarships and other grants to individuals, and employee matching gifts and excludes set-asides, loans, program-related investments (PRIs), and program expenses (see "Measures of Foundation Support, 2003").
2. To avoid risk or for lack of professional management, smaller endowed foundations tend to invest less heavily in the stock market and more in fixed income assets, including bonds and money market funds. Larger foundations invest more in stocks, but their investments are also more

diversified and may not match yields reported by the fastest growing stock indexes. In the *Foundation Management Report*, 10th edition, Council on Foundations, 2001, Table 2.1, the report's authors note that based on the latest member survey, close to two-thirds (65 percent) of the average foundation portfolio is composed of stocks and nearly one-fourth (24 percent) is made up of fixed income assets.

CHAPTER 3

Foundations by Region and State

Regional Trends, 1975 to 2003

Since the earliest days of philanthropy, the Northeast has held a disproportionate share of foundations and their resources. Beginning in the 1970s and accelerating in the 1980s, however, population shifts and strong economic growth in the South and West altered the philanthropic map. During the first half of the 1990s, change was more gradual. But the rapid growth of foundations in the West during the latter part of the decade and the above-average growth in the South's foundation community contributed to a more even distribution of foundation resources across regions. The following analysis examines changes in foundation number, giving, and assets over the latest year and since 1975 that have led to the current allocation of foundation resources across the four major U.S. regions (Table 9).

Foundation Number by Region

- West experienced fastest rate of growth in number of foundations in 2003

- Northeast continued to lead in terms of total foundation number

Growth of Number of Foundations, 1975 to 2003. The number of U.S. grantmaking foundations grew 2.4 percent in 2003 (Figure 13), roughly half the prior year's growth rate of 4.9 percent. The rate of growth has gradually slowed since 2000's record 12.7 percent gain. The West reclaimed its top spot based on growth in number of active grantmakers, reporting a 3.4 percent gain. The South slid to second place with a growth rate of 3 percent. The foundation communities of the Northeast and Midwest grew at a slower pace— 2.3 percent and 1.2 percent, respectively.

The strong performance of the West is consistent with its robust record of growth for almost the past two decades. Figure 14 shows that the number of active foundations in the region has grown more than fourfold (321.4 percent) since 1975, surpassing the increase for all U.S. foundations (203.4 percent). The South recorded the second fastest growth rate during this period (279.7 percent), followed by the Midwest (175.3 percent) and the Northeast (144.1 percent). However, despite the relatively slow growth rate recorded by the Northeast for the period between 1975 and 2003, the region ranked first by the total number of active foundations in the latest year, reflecting the presence of a strong foundation base prior to 1975.

Growth of Actual Number of Foundations. An examination of the growth rate of a foundation community provides a useful measure of the strength of philanthropy in a region. However, regions with larger and more established foundation communities, such as the Northeast and Midwest, may experience increases in actual foundation resources that surpass faster-growing regions with a smaller philanthropic base. For example, Figure 15 shows that, while the Northeast experienced the slowest rate of growth in number of active foundations between 1975 and 2003, this region still accounted for the second largest number of new foundations in this period (12,278). By comparison, the rapidly growing West reported the lowest number of actual foundations added (8,846). Interestingly, the South, which began with a relatively smaller foundation base, surpassed the Northeast in actual number of additional foundations counted (12,839). In the latest year alone, the South reported an increase of 505 active foundations, bringing the regional total to 17,430 (Table 9). The Northeast recorded the second largest increase in actual number of active foundations in 2003 (474), followed by the West (378) and Midwest (196) regions.

TABLE 9. Fiscal Data of Grantmaking Foundations by Region and State, 2003

Region[1]	Number of Foundations	%	Assets	%	Gifts Received	%	Qualifying Distributions[2]	%	Total Giving[3]	%
NORTHEAST	**20,797**	**31.3**	**$142,893,232**	**30.0**	**$ 6,784,798**	**27.3**	**$11,505,084**	**33.9**	**$10,564,620**	**34.9**
New England	**6,008**	**9.0**	**22,415,629**	**4.7**	**1,316,580**	**5.3**	**1,762,580**	**5.2**	**1,614,574**	**5.3**
Connecticut	1,369	2.1	5,915,114	1.2	335,633	1.4	537,019	1.6	514,978	1.7
Maine	272	0.4	934,266	0.2	79,213	0.3	68,655	0.2	61,881	0.2
Massachusetts	2,753	4.1	11,810,978	2.5	733,301	2.9	907,522	2.7	824,520	2.7
New Hampshire	280	0.4	985,622	0.2	71,612	0.3	67,649	0.2	49,923	0.2
Rhode Island	1,111	1.7	2,376,179	0.5	64,902	0.3	154,315	0.5	143,468	0.5
Vermont	223	0.3	393,470	0.1	31,919	0.1	27,420	0.1	19,804	0.1
Middle Atlantic	**14,789**	**22.3**	**120,477,603**	**25.3**	**5,468,218**	**22.0**	**9,742,503**	**28.7**	**8,950,046**	**29.5**
New Jersey	2,436	3.7	16,994,613	3.6	1,745,451	7.0	2,129,901	6.3	1,969,120	6.5
New York	8,840	13.3	78,918,087	16.6	3,151,939	12.7	5,955,444	17.5	5,446,687	18.0
Pennsylvania	3,513	5.3	24,564,903	5.2	570,828	2.3	1,657,159	4.9	1,534,238	5.1
MIDWEST	**16,560**	**24.9**	**103,538,330**	**21.7**	**4,365,862**	**17.6**	**7,147,419**	**21.1**	**6,425,111**	**21.2**
East North Central	**11,829**	**17.8**	**78,941,411**	**16.6**	**3,227,740**	**13.0**	**5,316,899**	**15.7**	**4,743,711**	**15.7**
Illinois	3,807	5.7	20,936,489	4.4	1,185,237	4.8	1,384,608	4.1	1,270,086	4.2
Indiana	1,131	1.7	17,163,107	3.6	294,616	1.2	841,212	2.5	801,082	2.6
Michigan	1,923	2.9	21,611,596	4.5	524,057	2.1	1,306,418	3.8	1,211,590	4.0
Ohio	3,019	4.5	13,222,609	2.8	881,071	3.5	1,297,593	3.8	1,002,104	3.3
Wisconsin	1,949	2.9	6,007,611	1.3	342,759	1.4	487,068	1.4	458,849	1.5
West North Central	**4,731**	**7.1**	**24,596,919**	**5.2**	**1,138,122**	**4.6**	**1,830,519**	**5.4**	**1,681,400**	**5.5**
Iowa	808	1.2	2,298,452	0.5	162,020	0.7	204,856	0.6	153,718	0.5
Kansas	666	1.0	1,829,493	0.4	69,397	0.3	117,079	0.3	105,339	0.3
Minnesota	1,295	2.0	10,162,212	2.1	329,125	1.3	728,528	2.1	674,001	2.2
Missouri	1,236	1.9	7,842,964	1.6	362,072	1.5	548,414	1.6	524,296	1.7
Nebraska	523	0.8	1,966,156	0.4	175,006	0.7	200,957	0.6	196,748	0.6
North Dakota	80	0.1	160,470	0.0	9,870	0.0	9,022	0.0	7,224	0.0
South Dakota	123	0.2	337,173	0.1	30,633	0.1	21,663	0.1	20,074	0.1
SOUTH	**17,430**	**26.3**	**101,137,232**	**21.2**	**6,300,512**	**25.3**	**7,557,336**	**22.3**	**6,679,543**	**22.0**
South Atlantic	**10,586**	**15.9**	**58,166,270**	**12.2**	**3,030,667**	**12.2**	**4,573,819**	**13.5**	**4,016,548**	**13.3**
Delaware	331	0.5	2,992,390	0.6	88,568	0.4	203,468	0.6	195,957	0.6
District of Columbia	383	0.6	4,265,512	0.9	355,164	1.4	512,991	1.5	358,324	1.2
Florida	3,390	5.1	14,027,413	2.9	661,385	2.7	1,069,301	3.2	975,233	3.2
Georgia	1,276	1.9	8,990,413	1.9	476,026	1.9	765,939	2.3	647,630	2.1
Maryland	1,381	2.1	10,373,832	2.2	303,409	1.2	705,480	2.1	637,359	2.1
North Carolina	2,014	3.0	9,576,538	2.0	532,815	2.1	723,598	2.1	690,432	2.3
South Carolina	377	0.6	1,421,497	0.3	61,239	0.2	91,081	0.3	81,968	0.3
Virginia	1,208	1.8	5,767,406	1.2	507,501	2.0	462,935	1.4	393,847	1.3
West Virginia	226	0.3	751,269	0.2	44,561	0.2	39,028	0.1	35,796	0.1
East South Central	**2,002**	**3.0**	**8,111,101**	**1.7**	**584,233**	**2.4**	**660,724**	**1.9**	**602,839**	**2.0**
Alabama	660	1.0	1,851,327	0.4	74,130	0.3	154,017	0.5	144,987	0.5
Kentucky	416	0.6	1,503,373	0.3	104,267	0.4	89,518	0.3	83,344	0.3
Mississippi	231	0.3	820,615	0.2	44,145	0.2	59,183	0.2	52,401	0.2
Tennessee	695	1.0	3,935,787	0.8	361,690	1.5	358,006	1.1	322,106	1.1
West South Central	**4,842**	**7.3**	**34,859,861**	**7.3**	**2,685,611**	**10.8**	**2,322,794**	**6.8**	**2,060,157**	**6.8**
Arkansas	255	0.4	1,868,087	0.4	162,931	0.7	277,320	0.8	272,106	0.9
Louisiana	440	0.7	1,849,500	0.4	87,693	0.4	138,435	0.4	123,347	0.4
Oklahoma	579	0.9	6,122,211	1.3	706,693	2.8	291,380	0.9	214,785	0.7
Texas	3,568	5.4	25,020,063	5.2	1,728,294	7.0	1,615,658	4.8	1,449,918	4.8
WEST	**11,598**	**17.5**	**129,083,591**	**27.1**	**7,402,100**	**29.8**	**7,726,859**	**22.8**	**6,635,673**	**21.9**
Mountain	**3,219**	**4.8**	**17,843,789**	**3.7**	**783,290**	**3.2**	**1,073,870**	**3.2**	**1,000,779**	**3.3**
Arizona	538	0.8	2,438,581	0.5	103,879	0.4	160,308	0.5	142,972	0.5
Colorado	1,105	1.7	6,739,856	1.4	337,878	1.4	361,031	1.1	321,701	1.1
Idaho	186	0.3	1,018,761	0.2	83,016	0.3	49,215	0.1	45,787	0.2
Montana	177	0.3	357,657	0.1	14,279	0.1	20,420	0.1	18,790	0.1
Nevada	426	0.6	3,479,143	0.7	103,848	0.4	215,445	0.6	233,356	0.8
New Mexico	203	0.3	1,123,984	0.2	56,951	0.2	72,217	0.2	55,169	0.2
Utah	405	0.6	1,790,690	0.4	43,263	0.2	132,367	0.4	126,590	0.4
Wyoming	179	0.3	895,117	0.2	40,175	0.2	62,865	0.2	56,414	0.2
Pacific	**8,379**	**12.6**	**111,239,802**	**23.3**	**6,618,810**	**26.6**	**6,652,990**	**19.6**	**5,634,894**	**18.6**
Alaska	62	0.1	301,875	0.1	34,439	0.1	18,591	0.1	15,915	0.1
California	6,160	9.3	71,862,374	15.1	5,941,471	23.9	4,503,866	13.3	3,759,678	12.4
Hawaii	272	0.4	1,590,056	0.3	45,874	0.2	94,797	0.3	72,608	0.2
Oregon	677	1.0	3,475,625	0.7	296,658	1.2	240,452	0.7	215,393	0.7
Washington	1,208	1.8	34,009,872	7.1	300,367	1.2	1,795,283	5.3	1,571,299	5.2
CARIBBEAN[4]	**12**	**0.0**	**59,418**	**0.0**	**5,054**	**0.0**	**4,352**	**0.0**	**3,840**	**0.0**
Puerto Rico	6	0.0	30,790	0.0	1,764	0.0	1,475	0.0	1,443	0.0
Virgin Islands	6	0.0	28,628	0.0	3,290	0.0	2877	0.0	2,397	0.0
SOUTH PACIFIC	**1**	**0.0**	**1,311**	**0.0**	**0**	**0.0**	**48**	**0.0**	**48**	**0.0**
American Samoa	1	0.0	1,311	0.0	0	0.0	48	0.0	48	0.0
Total	**66,398**	**100.0**	**$476,713,115**	**100.0**	**$24,858,326**	**100.0**	**$33,941,098**	**100.0**	**$30,308,835**	**100.0**

Source: The Foundation Center, *Foundation Yearbook,* 2005. Dollars in thousands. Due to rounding, figures may not add up.
[1]Geographic regions as defined by the U.S. Census Bureau.
[2]Qualifying distributions are the expenditures used in calculating the required payout; includes total giving, as well as reasonable administrative expenses, set-asides, PRIs, operating program expenses, and amount paid to acquire assets used directly for charitable purposes.
[3]Includes grants, scholarships, and employee matching gifts; excludes set-asides, loans, PRIs, and program expenses. For some operating foundations, program expenses are included.
[4]Private foundations in Puerto Rico, the Virgin Islands and American Samoa are not required to file Form 990-PF. Only a few voluntary reporters are represented.

Giving by Region

• **Northeast experienced the fastest growth in giving in 2003**

• **West surpassed Midwest by share of overall giving for the first time on record**

Change in Giving. Overall foundation giving decreased 0.4 percent in 2003, following a 0.2 percent dip in the preceding year (Figure 13). In terms of actual dollar value, giving slipped from $30.4 billion in 2002 to $30.3 billion in the latest year. Nonetheless, two of the four regions experienced modest growth in giving. Grantmakers in the Northeast showed the largest gain in total giving (3.1 percent), followed by the West (1.3 percent). In contrast, foundations in the Midwest and the South reported declines in giving in 2003 (5.4 percent and 2.4 percent, respectively).

FIGURE 13.   Change in Foundation Number, Giving, and Assets by Region, 2002 to 2003

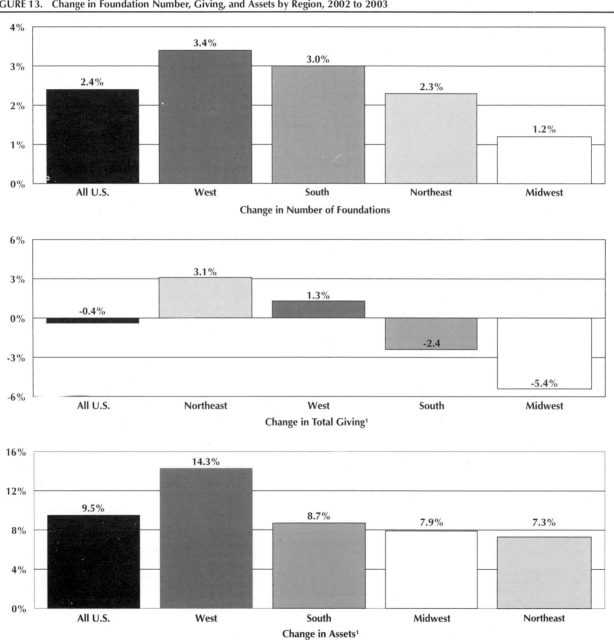

Source: The Foundation Center, *Foundation Yearbook,* 2005.
[1]Percent change based on current dollars.

Change in Actual Grant Dollars. In addition to the fastest rate of growth in giving in 2003, the Northeast also reported the largest increase in terms of actual grant dollars. Grantmakers in the Northeast raised their giving by almost $320 million to $10.6 billion in the latest year. New Jersey, Massachusetts, and Connecticut posted the largest gains. Foundations in the West ranked second, with giving up by $84.7 million to over $6.6 billion. Meanwhile, giving in the South decreased by $164.3 million to $6.7 billion and Midwestern foundations decreased giving by $364 million to $6.4 billion.

Change in Giving, 1975 to 2003. Since 1975, the West has led in the growth of foundation giving (Figure 16). Grantmaking by foundations in the region has increased over elevenfold (1,031.4 percent) after inflation during this period, far surpassing growth reported by foundations in the South (494.2 percent) and U.S. foundations overall (355.7 percent). With close to a fourfold increase in giving since the mid-1970s (269.7 percent), funders based in the Midwest fell behind the average growth in U.S. foundation giving. Support from grantmakers in the Northeast grew more slowly (229.9 percent).

As a result of faster growth in foundation giving in regions with a smaller philanthropic base, grantmaking has become far more evenly distributed across regions. Figure 17 shows that in 1975 foundations in the South accounted for about one-in-six grant dollars (16.9 percent) awarded by U.S. foundations. In the latest year, they accounted for more than one-fifth (22 percent) of overall grant dollars in the United States.

Among grantmakers in the West, their share of overall grant dollars has risen from less than one-in-eleven (8.8 percent) to more than one-in-five (21.9 percent). With signs of recovery in the technology sector (an important source of support for the West's foundation community), prospects for continued growth in giving by Western foundations remain strong. The region surpassed the Midwest in share of overall foundation giving for the first time in 2003 and will likely surpass the South in the next few years.

In contrast to rapid growth in the South and West, the Midwest has maintained a roughly one-quarter

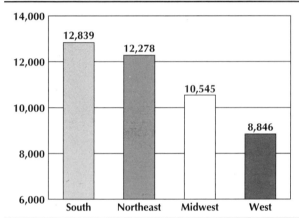

FIGURE 15.   Change in Number of Grantmaking Foundations by Region, 1975 to 2003

Source: The Foundation Center, *Foundation Yearbook,* 2005. Overall, the Northeast reported 20,797 active foundations in 2003, followed by the South (17,430), Midwest (16,560), and West (11,598).

FIGURE 14.   Change in Foundation Number by Region, 1975 to 2003

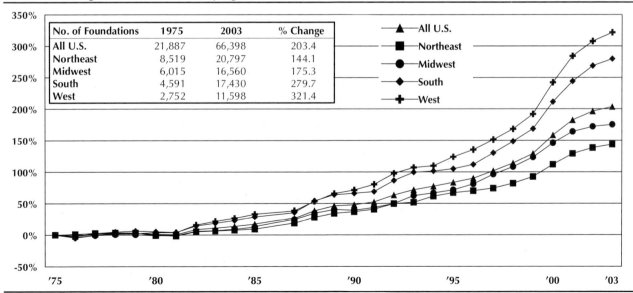

No. of Foundations	1975	2003	% Change
All U.S.	21,887	66,398	203.4
Northeast	8,519	20,797	144.1
Midwest	6,015	16,560	175.3
South	4,591	17,430	279.7
West	2,752	11,598	321.4

Source: The Foundation Center, *Foundation Yearbook,* 2005. Figures for 1986 not available.

share of giving over the past quarter century, though in recent years that share has fallen closer to one-fifth (21.2 percent). Northeastern foundations' share of giving, however, has declined from close to half (48.1 percent) to just over one-third (34.9 percent) during this period. Nonetheless, funders in the region continued to account for the largest share of total grant dollars. In general, growth in giving has been strong across the United States since 1975, with the benefits of this increased support reaching many more parts of the country.

Assets by Region

• **U.S. foundation assets increased across all regions**

• **West experienced largest gain in assets in 2003**

Change in Assets. U.S. foundation assets increased 9.5 percent in 2003, after two consecutive years of decline (Figure 13). This growth reflected the stock market recovery and improvement in the economy, which boosted existing foundation assets and encouraged donors to increase their new gifts into foundations. Grantmakers in the West contributed heavily to the increase in U.S. foundation assets, with an asset gain of 14.3 percent. Within this region, California recorded the highest increase (19.4 percent), followed by Oregon (13.3 percent) and Colorado (12.8 percent). Foundations in the South trailed those in the West,

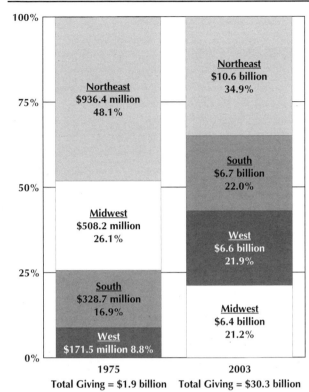

FIGURE 17. Foundation Giving by Region, 1975 and 2003

1975 — Total Giving = $1.9 billion
2003 — Total Giving = $30.3 billion

Source: The Foundation Center, *Foundation Yearbook,* 2005. Figures based on unadjusted dollars. Due to rounding, percentages may not total 100. Total giving figure includes foundations based in the Caribbean and South Pacific, which accounted for 0.0 percent of total foundation giving in 1975 and 2003.

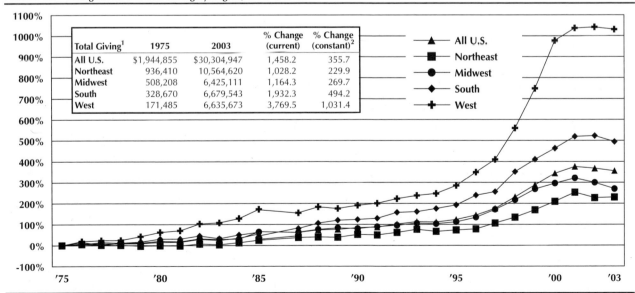

FIGURE 16. Change in Foundation Giving by Region, 1975 to 2003

Total Giving[1]	1975	2003	% Change (current)	% Change (constant)[2]
All U.S.	$1,944,855	$30,304,947	1,458.2	355.7
Northeast	936,410	10,564,620	1,028.2	229.9
Midwest	508,208	6,425,111	1,164.3	269.7
South	328,670	6,679,543	1,932.3	494.2
West	171,485	6,635,673	3,769.5	1,031.4

Source: The Foundation Center, *Foundation Yearbook,* 2005. Figures for 1986 not available.
[1]Dollar figures in thousands. Figures based on current dollars.
[2]Constant 1975 dollars based on annual average Consumer Price Index, all urban consumers, as reported by the U.S. Department of Labor, Bureau of Labor Statistics, as of February 2005.

posting asset growth of 8.7 percent. The Midwest and the Northeast, with more established philanthropic bases, ranked third and fourth and recorded gains of 7.9 percent and 7.3 percent, respectively.

Differences in asset increases across regions partly depended on the influx of new gifts into foundations. For example, the two top regions in terms of growth in assets—the West and the South—reported substantial increases in gifts received (55.5 percent and 23.2 percent, respectively). By comparison, the Midwest and the Northeast reported a more modest growth in new gifts in 2003.

Change in Actual Asset Dollars. By dollar amount, the increase in foundation assets in the West in 2003 far exceeded the gains reported for other regions. Overall assets of foundations in the West rose by $16.2 billion to $129.1 billion.

Northeastern foundations experienced the second largest asset gain in 2003—$9.7 billion—raising their total assets to $142.9 billion. Assets of foundations in the South increased by roughly $8.1 billion to $101.1 billion. Finally, grantmakers in the Midwest realized a $7.5 billion gain in their asset value, taking it to $103.5 billion.

Change in Assets, 1975 to 2003. A historical review of trends in foundation assets since 1975 reveals that all regions experienced substantial gains in asset levels over the past three decades. Among the four major regions, the West showed the fastest growth (Figure 18). Adjusted for inflation, assets of foundations in the West climbed close to fifteenfold (1,365.8 percent) from 1975 to 2003—providing the basis for the region's elevenfold (1,031.4 percent) increase in giving.

A substantial share of the asset growth in the West over the last quarter century came from bequests received in the late 1970s or early 1980s, such as the endowment of the J. Paul Getty Trust in California. During the late 1990s, California experienced a major burst of new asset growth from both established and new foundations. First, the David and Lucile Packard Foundation received its principal endowment from its founders in two bequests valued at $5.7 billion. Second, the California Endowment, a new health foundation, was formed with an initial endowment of $1.3 billion. Moreover, from the late 1990s to 2001, the Bill & Melinda Gates Foundation (established in 1994 as the William H. Gates Foundation) received several multi-billion-dollar annual gifts from Bill & Melinda

FIGURE 18.   Change in Foundation Assets by Region, 1975 to 2003

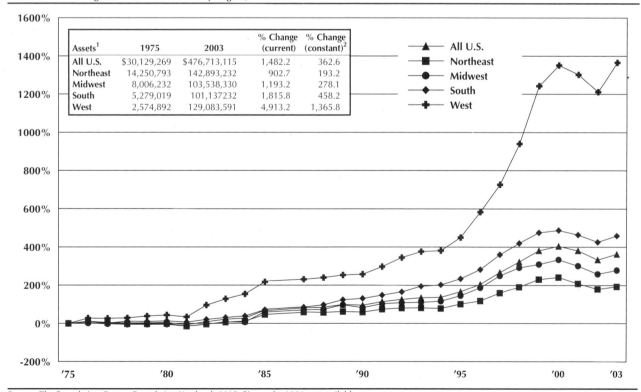

Assets[1]	1975	2003	% Change (current)	% Change (constant)[2]
All U.S.	$30,129,269	$476,713,115	1,482.2	362.6
Northeast	14,250,793	142,893,232	902.7	193.2
Midwest	8,006,232	103,538,330	1,193.2	278.1
South	5,279,019	101,137232	1,815.8	458.2
West	2,574,892	129,083,591	4,913.2	1,365.8

Source: The Foundation Center, *Foundation Yearbook,* 2005. Figures for 1986 not available.
[1]Dollar figures in thousands. Figures based on current dollars.
[2]Constant 1975 dollars based on annual average Consumer Price Index, all urban customers, as reported by U.S. Department of Labor, Bureaus of Labor Statistics, as of February 2005.

Gates totaling $23.5 billion. A $3.1 billion bequest from the estate of Hewlett-Packard Company co-founder Bill Hewlett to the California-based Hewlett Foundation in 2001 also helped to boost assets. Finally, a $3.7 billion gift in 2003 from Intel Corporation co-founder, Gordon E. Moore, and his wife, Betty, to their California foundation added to the assets in the West.

Grantmakers in the South also posted above-average asset growth since 1975. Over the past quarter century, the real value of assets of foundations based in the South rose well over fivefold (458.2 percent), surpassing the national average (362.6 percent). Foundation growth in this region accelerated in the mid-1980s along with regional economic growth and continued into the 1990s. Several established foundations received substantial additions to their endowments (e.g., Burroughs Wellcome Fund and Walton Family Foundation), large new foundations were formed (e.g., Freedom Forum, Goizueta Foundation, and Michael and Susan Dell Foundation), a few leading foundations relocated to the region (e.g., Arthur S. DeMoss Foundation and John S. and James L. Knight Foundation), and corporate foundation activity increased.

FIGURE 19. Foundation Assets by Region, 1975 and 2003

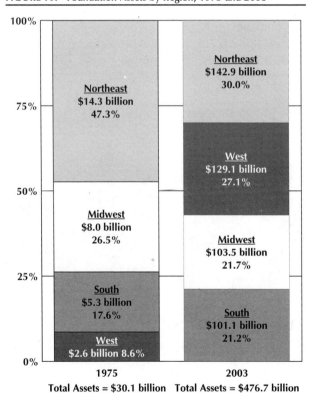

Total Assets = $30.1 billion   Total Assets = $476.7 billion

Source: The Foundation Center, *Foundation Yearbook,* 2005. Figures based on unadjusted dollars. Total assets figure includes assets of foundations based in the Caribbean and South Pacific, which accounted for 0.0 percent of total foundation assets in 1975 and 2003.

As of 2003, the South claimed ten of the nation's 46 "billion-dollar" foundations, led by the Maryland-based Annie E. Casey Foundation.

Finally, inflation-adjusted asset growth of grantmakers in the Midwest (278.1 percent) and Northeast (193.2 percent) fell well below the average increase in overall U.S. foundation assets since 1975.

The significant growth of foundation assets in the West has led to pronounced changes in the distribution of foundation resources by region. In 1999, for example, the West surpassed the South and the Midwest in share of all foundation assets for the first time. Overall, between 1975 and 2003, the West more than tripled its share of foundation assets from 8.6 percent to 27.1 percent (Figure 19).

Following the West, the South increased its share of all foundation assets from over one-sixth (17.6 percent) in 1975 to more than one-fifth (21.2 percent) in 2003. In contrast, the Midwest slipped from over one-fourth (26.5 percent) to more than one-fifth (21.7 percent) of U.S. foundation assets, while the Northeast region dropped from close to half (47.3 percent) to less than one-third (30 percent) of assets.

Nonetheless, all regions benefited from robust growth in foundation assets since the mid-1970s, and grantmakers in the Northeast continued to control by far the largest share of foundation resources.

State-Level Trends, 1975 to 2003

The preceding analysis examined variations in the growth of foundation number, giving, and assets for the four major regions of the United States. The following section examines notable differences in the growth and distribution of foundation resources among individual states. As Figure 20 shows, the number of foundations has grown dramatically across the United States. In 1975, only one state (New York) registered 2,000 or more foundations. By 2003, ten states (New York, California, Illinois, Texas, Pennsylvania, Florida, Ohio, Massachusetts, New Jersey, and North Carolina) reported having that many foundations.

Giving by State

• **Maine experienced most rapid rate of growth in giving in 2003**

• **New Jersey led in actual grant dollar gain**

Change in Giving. Nationally, foundation giving fell 0.4 percent between 2002 and 2003. At the state level, 16 states plus the District of Columbia reported a gain in giving. Maine showed the fastest growth

(30.1 percent), raising its grant dollars from $47.6 million to $61.9 million (Figure 21). The Libra Foundation, with giving of $16.1 million in 2003, accounted for most of this growth. Arkansas followed with a 24.3 percent rise in grants dollars, from about $2.2 million to $2.7 million. New Jersey took the third spot with a 22.1 percent increase in giving, from $1.6 billion to close to $2 billion.

In contrast, 34 states showed decreases in foundation giving in 2003. Mississippi reported the greatest decline, with support down 57.2 percent to $52.4 million. This drop followed a more than doubling of its grant dollars in 2002. Idaho posted the second largest decrease in giving in 2003, with grant dollars down 28 percent to $45.8 million. Third among states showing the largest declines in giving was Louisiana, which reduced funding by 22.4 percent to $123.3 million.

Change in Actual Grant Dollars. As noted in the analysis by region, states with larger or more established foundation communities may not show as rapid growth in foundation resources. Yet these states may exceed faster growing states in terms of the increase in actual grant dollars. For example, while Maine reported the fastest rate of growth in foundation giving between 2002 and 2003, it ranked eighth in terms of gain in actual grant dollars. The state showing the largest increase in actual grant dollars in the latest year was New Jersey (up $356.6 million), although it ranked third by rate of growth. Much of the state's $356.6 million gain in giving was due to substantial increases in giving by three large operating foundations—the Merck

Patient Assistance Program, Bristol-Myers Squibb Patient Assistance Foundation, and Janssen Ortho Patient Assistance Foundation (see Table 4 in Chapter 2). Following New Jersey in the growth of actual grant dollars were California (up $148.5 million), Arkansas (up $53.2 million) and Massachusetts (up $49.3 million).

Giving: Average, Per Capita, and as a Share of GSP. Average giving by all 66,398 active U.S. foundations equaled $456,472 in 2003. Only 13 states plus the District of Columbia reported higher levels of average foundation giving. Washington State ($1,300,744) showed the highest average foundation giving in the latest year, followed by Arkansas ($1,067,082) and the District of Columbia ($935,972). These were also the top three states by average giving in 2002. Still, the overall number of foundations in each of these states tended to be small, and a single or small number of relatively large foundations accounted for the vast majority of giving. By comparison, New York, which is home to more than 8,800 foundations, albeit 10 of the 50 largest foundations by total giving, showed lower average giving of $616,141.

Other comparative measures of giving across states take into account socioeconomic factors such as state population and Gross State Product (GSP). Rankings of giving based on these criteria show that New York foundations ranked first by overall giving and giving as a percentage of GSP and second based on giving per capita (Table 10). Funders in the District of Columbia ranked first based on giving per capita. This may be due to the small size of the DC population relative to

FIGURE 20. Grantmaking Foundations by State, 1975 and 2003

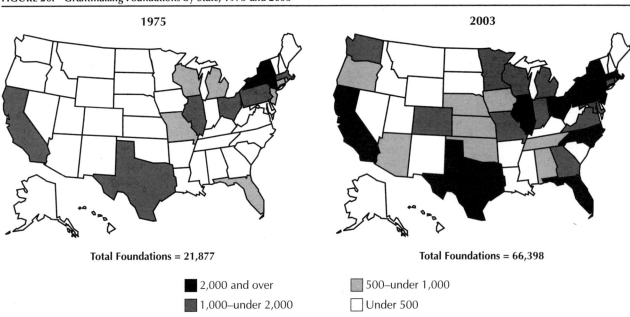

1975

2003

Total Foundations = 21,877

Total Foundations = 66,398

- ■ 2,000 and over
- ■ 1,000–under 2,000
- ▨ 500–under 1,000
- ☐ Under 500

Source: The Foundation Center, *Foundation Yearbook,* 2005.

other metropolitan areas and to the presence of a few large national and international funders in the area.

Distribution of Giving by State, 1975 and 2003. Varying patterns of foundation development across the United States have resulted in dramatic changes in the regional and state-level distribution of foundation resources. For example, while New York ranked first by foundation giving in both 1975 and 2003, Figure 22 shows that its share of overall grant dollars declined from roughly one-in-three (32.1 percent) to under one-in-five (18 percent). In contrast, California increased its share of overall U.S. foundation grant dollars during this period from roughly one-in-seventeen (5.8 percent) to nearly one-in-eight (12.4 percent). Still, just five states accounted for close to half (47.1 percent) of all foundation giving in 2003, although this share was down from the close to three-fifths (57.5 percent) of giving reported in 1975.

An examination of the ten largest states by foundation giving in 2003 showed that, with the exception of Washington State and Florida, all ranked among the top ten in 1975 (Table 11). Moreover, grantmakers in just those ten states provided two-thirds (66.6 percent) of grant dollars awarded by U.S. foundations overall.

Assets by State

- **Delaware reported highest rate of asset growth in 2003**

- **California topped all states in actual asset dollar gain**

In contrast to the analysis of changes in foundation giving by region, most states showed an increase in the value of their foundations' endowments over the past year. Still, considerable variation occurred among the states in the patterns of change in giving and assets.

Change in Assets. Nationally, foundation assets increased 9.5 percent between 2002 and 2003. All but four states posted increases in endowment values. In contrast, 40 states and DC reported declines in assets in 2002. Delaware registered the biggest percentage gain in the latest year, with foundation assets up 23.4 percent to almost $3 billion (Figure 23 and Table 9). Contributing to this growth were asset gains by several grantmakers, including the Delaware Community Foundation, whose assets rose 85.4 percent to $124.6 million.

Maine reported the second largest increase in foundation assets in 2003 (up 21 percent to $934.3 million). A number of multi-million-dollar gains accounted for this growth, such as the $43.5 million increase in assets for the Maine Community Foundation, the $25.3 million gain for the Libra Foundation, and the Maine Health Access Foundation's $16.2 million increase. Of the 47 states (including DC) that experienced asset gains, two states reported asset increases over 20 percent, 16 states reported asset gains of between 10 and 20 percent, 21 states posted increases between 5 and 10 percent, and eight states had increases of less than 5 percent.

Among the four states in which foundation assets declined in the latest year, Louisiana reported the largest decrease. Foundation assets in the state registered a 7.3 percent decline from almost $2 billion to $1.8 billion (Table 9). The other three states posted more modest reductions in foundation assets, ranging from 5.1 percent for Wyoming to 0.6 percent for Virginia.

FIGURE 21. Top Ten States by Change in Foundation Giving, 2002 to 2003

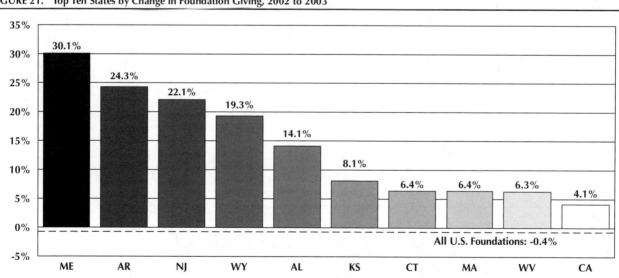

Source: The Foundation Center, *Foundation Yearbook,* 2005. Percent change based on current dollars.

TABLE 10. Foundation Giving Per Capita and as a Share of Gross State Product, 2003

Region[1]	Total Giving[2]	Rank	Giving Per Capita[3]	Rank	Giving as a % of Gross State Product[4]	Rank
NORTHEAST	**$10,564,620**		**$194**		**0.46**	
New England	**1,614,574**		**114**		**0.26**	
Connecticut	514,978	18	148	6	0.30	14
Maine	61,881	40	47	35	0.15	33
Massachusetts	824,520	11	128	10	0.28	15
New Hampshire	49,923	44	39	36	0.10	39
Rhode Island	143,468	32	133	7	0.36	7
Vermont	19,804	48	32	40	0.10	41
Middle Atlantic	**8,950,046**		**222**		**0.54**	
New Jersey	1,969,120	3	228	5	0.50	4
New York	5,446,687	1	284	2	0.66	1
Pennsylvania	1,534,238	5	124	11	0.34	9
MIDWEST	**6,425,111**		**98**		**0.27**	
East North Central	**4,743,711**		**103**		**0.28**	
Illinois	1,270,086	7	100	18	0.25	19
Indiana	801,082	12	129	9	0.37	6
Michigan	1,211,590	8	120	12	0.33	10
Ohio	1,002,104	9	88	21	0.25	21
Wisconsin	458,849	19	84	22	0.23	22
West North Central	**1,681,400**		**86**		**0.23**	
Iowa	153,718	30	52	34	0.15	34
Kansas	105,339	36	39	37	0.11	37
Minnesota	674,001	14	133	8	0.32	11
Missouri	524,296	17	92	20	0.27	16
Nebraska	196,748	28	113	14	0.30	13
North Dakota	7,224	51	11	51	0.03	51
South Dakota	20,074	47	26	43	0.08	45
SOUTH	**6,679,543**		**64**		**0.18**	
South Atlantic	**4,016,548**		**74**		**0.20**	
Delaware	195,957	29	240	4	0.40	5
District of Columbia	358,324	21	642	1	0.51	3
Florida	975,233	10	57	30	0.18	28
Georgia	647,630	15	75	24	0.20	25
Maryland	637,359	16	116	13	0.30	12
North Carolina	690,432	13	82	23	0.22	23
South Carolina	81,968	38	20	49	0.06	49
Virginia	393,847	20	53	33	0.13	35
West Virginia	35,796	46	20	48	0.08	44
East South Central	**602,839**		**35**		**0.11**	
Alabama	144,987	31	32	39	0.11	38
Kentucky	83,344	37	20	47	0.06	48
Mississippi	52,401	43	18	50	0.07	47
Tennessee	322,106	22	55	31	0.16	31
West South Central	**2,060,157**		**63**		**0.18**	
Arkansas	272,106	24	100	19	0.36	8
Louisiana	123,347	35	27	42	0.09	42
Oklahoma	214,785	27	61	27	0.21	24
Texas	1,449,918	6	66	26	0.18	27
WEST	**6,635,673**		**100**		**0.26**	
Mountain	**1,000,779**		**52**		**0.15**	
Arizona	142,972	33	26	44	0.08	43
Colorado	321,701	23	71	25	0.17	29
Idaho	45,787	45	33	38	0.11	36
Montana	18,790	49	20	46	0.07	46
Nevada	233,356	25	104	17	0.27	17
New Mexico	55,169	42	29	41	0.10	40
Utah	126,590	34	54	32	0.17	30
Wyoming	56,414	41	112	15	0.25	20
Pacific	**5,634,894**		**120**		**0.30**	
Alaska	15,915	50	25	45	0.05	50
California	3,759,678	2	106	16	0.26	18
Hawaii	72,608	39	58	29	0.15	32
Oregon	215,393	26	60	28	0.18	26
Washington	1,571,299	4	256	2	0.64	2
TOTAL[5]	**$30,308,835**		**$104**		**0.28**	

Source: The Foundation Center, *Foundation Yearbook,* 2005.
[1]Geographic regions as defined by the U.S. Census Bureau.
[2]Total giving figures in thousands. Due to rounding, figures may not add up.
[3]Per capita giving figures in actual dollar amounts. Based on July 1, 2003 population estimates (U.S. Census Bureau).
[4]Based on 2003 Gross State Product figures (U.S. Department of Commerce, Bureau of Economic Analysis, Regional Economic Analysis Division, December 2004).
[5]Includes giving by 13 foundations based in the Caribbean and South Pacific.

Change in Actual Asset Dollars. Among the top ten states by foundation assets overall, eight also ranked among the largest by increases in actual asset values in 2003. California led with a total asset gain of $11.7 billion in the latest year. Three foundations accounted for over one-fourth of this increase: the William and Flora Hewlett and the David and Lucile Packard foundations, which each reported asset gains of over $1.8 billion, and the California Endowment, which showed an increase of over $800 million.

New York registered the second largest gain in actual asset values in 2003, increasing over $5 billion. More than one-third of this gain was attributable to the Andrew W. Mellon Foundation, with assets up by $1.1 billion, and the Ford Foundation, with assets up by $670.6 million. By comparison, in the previous year,

New York and California incurred the biggest losses in terms of actual asset value.

Aside from California and New York, eight other states posted asset increases of over $1 billion: Washington State, with foundation assets up $3 billion; Michigan, with assets up $2.4 billion; Texas, with assets up $2.3 billion; Pennsylvania, with assets up just over $2 billion; Indiana, with assets up $1.5 billion; Minnesota, with assets up nearly $1.3 billion; and New Jersey and Ohio, with assets up just over $1.2 billion and $1.1 billion, respectively.

Distribution of Assets by State, 1975 and 2003. Strong foundation growth outside of the more established Northeast and Midwest regions has resulted in pronounced shifts in foundation resources by state.

TABLE 11. Top Ten States by Total Giving Reported, 2003

State	No. of Foundations	%	Total Giving[1]	%	Average Giving Per Foundation	Rank 1975
1. New York	8,840	13.3	$ 5,446,687	18.0	$ 616	1
2. California	6,160	9.3	3,759,678	12.4	610	4
3. New Jersey	2,436	3.7	1,969,120	6.5	808	9
4. Washington	1,208	1.8	1,571,299	5.2	1,301	22
5. Pennsylvania	3,513	5.3	1,534,238	5.1	437	2
6. Texas	3,568	5.4	1,449,918	4.8	406	5
7. Illinois	3,807	5.7	1,270,086	4.2	334	6
8. Michigan	1,923	2.9	1,211,590	4.0	630	3
9. Ohio	3,019	4.5	1,002,104	3.3	332	7
10. Florida	3,390	5.1	975,233	3.2	288	18
Subtotal	**37,864**	**57.0**	**$20,189,953**	**66.6**	**$ 533**	
All other states	28,534	43.0	$10,118,882	33.4	$ 355	
Total	**66,398**	**100.0**	**$30,308,835**	**100.0**	**$ 456**	

Source: The Foundation Center, *Foundation Yearbook*, 2005. Dollars in thousands. Due to rounding, figures may not add up.

[1]Total giving equals total dollars paid out by foundations in each state and not dollars received by organizations in these states. Includes grants, scholarships, and employee matching gifts; excludes set-asides, loans, PRIs, and program expenses.

FIGURE 22. Distribution of Foundation Giving by State, 1975 and 2003

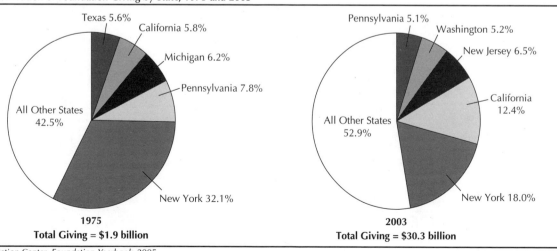

1975
Total Giving = $1.9 billion

2003
Total Giving = $30.3 billion

Source: The Foundation Center, *Foundation Yearbook*, 2005.

Figure 24 shows that, while New York ranked first by assets in both 1975 and 2003, its share of all asset dollars dropped from more than three-tenths (31.3 percent) to roughly one-sixth (16.6 percent). Conversely, California's share of all U.S. foundation assets close to tripled from one-eighteenth (5.5 percent) in 1975 to over one-seventh (15.1 percent) in 2003. Still, as was the case with overall giving, nearly half of foundation assets (49.2 percent) were concentrated in just five states, although this share was down from nearly three-fifths (58.3 percent) in 1975.

Regional and State-Level Distribution by Foundation Type, 2003

Just as overall foundation resources are spread unevenly across the four major regions of the United States, considerable regional variation exists in the distribution of foundations and their resources by foundation type. The following analysis provides a brief overview of the resources of independent foundations by region, followed by more detailed regional examinations of the resources of corporate and community foundations.

FIGURE 23.   Top Ten States by Change in Foundation Assets, 2002 to 2003

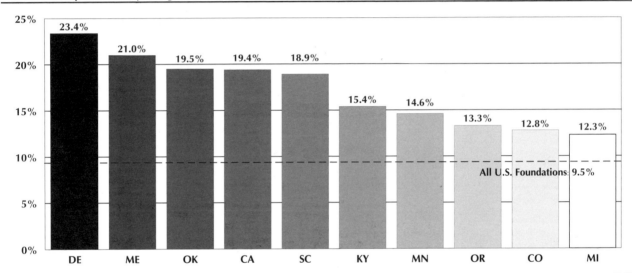

Source: The Foundation Center, *Foundation Yearbook*, 2005. Percent change based on current dollars.

FIGURE 24.   Distribution of Foundation Assets by State, 1975 and 2003

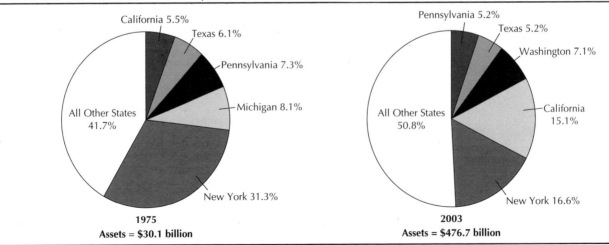

Source: The Foundation Center, *Foundation Yearbook*, 2005.

Independent Foundations

- **Independent foundations in the West accounted for largest share of giving by region in 2003**

Across regions, independent foundations by far represented the largest share of the U.S. foundation community. These funders, which include family and "new health" foundations, accounted for roughly nine out of ten foundations in each of the four major regions in 2003 (Table 12). Still, their share of grant dollars ranged from about seven-in-ten to more than eight-in-ten, with independents in the West reporting the largest share of giving (81.8 percent) (Figure 25). By assets, independent grantmakers in the Northeast held about nine-tenths (88.7 percent) of assets in the region, followed by independents in the South (82.8 percent), the West (81.2 percent), and the Midwest (80.9 percent).

Corporate Foundations

- **Midwest held largest number of corporate foundations in 2003**

- **Midwest accounted for largest share of corporate foundation giving by region**

- **New York led in number of corporate foundations by state**

More corporate foundations are located in the Midwest than in any other region. In 2003, over one-third (34.4 percent) of the 2,549 active U.S. business-sponsored foundations were based in the twelve-state region (Table 13). These funders also accounted for a larger share of overall giving within the Midwest region (17.2 percent) than did corporate foundations in the South (13.7 percent), the Northeast (10.7 percent), and the West (4.8 percent) (Figure 25).

TABLE 12.   Fiscal Data by Region and Foundation Type, 2003

NORTHEAST

Foundation Type	No. of Foundations	%	Assets	%	Gifts Received	%	Qualifying Distributions[1]	%	Total Giving[2]	%
Independent	18,755	90.2	$126,753,586	88.7	$3,722,168	54.9	$ 8,574,924	74.5	$ 7,798,064	73.8
Corporate	733	3.5	4,946,682	3.5	991,437	14.6	1,191,268	10.4	1,126,498	10.7
Community	95	0.5	6,568,256	4.6	447,717	6.6	422,867	3.7	424,919	4.0
Operating	1,214	5.8	4,624,708	3.2	1,623,476	23.9	1,316,025	11.4	1,215,138	11.5
Total	**20,797**	**100.0**	**$142,893,232**	**100.0**	**$6,784,798**	**100.0**	**$11,505,084**	**100.0**	**$10,564,620**	**100.0**

MIDWEST

Foundation Type	No. of Foundations	%	Assets	%	Gifts Received	%	Qualifying Distributions[1]	%	Total Giving[2]	%
Independent	14,537	87.8	$ 83,788,072	80.9	$2,258,654	51.7	$4,761,641	66.6	$4,441,358	69.1
Corporate	877	5.3	4,787,699	4.6	979,925	22.4	1,354,860	19.0	1,106,613	17.2
Community	336	2.0	12,603,728	12.2	995,331	22.8	791,910	11.1	791,538	12.3
Operating	810	4.9	2,358,831	2.3	131,952	3.0	239,007	3.3	85,601	1.3
Total	**16,560**	**100.0**	**$103,538,330**	**100.0**	**$4,365,862**	**100.0**	**$7,147,419**	**100.0**	**$6,425,111**	**100.0**

SOUTH

Foundation Type	No. of Foundations	%	Assets	%	Gifts Received	%	Qualifying Distributions[1]	%	Total Giving[2]	%
Independent	15,545	89.2	$ 83,704,373	82.8	$3,911,753	62.1	$5,331,725	70.6	$4,900,732	73.4
Corporate	586	3.4	4,040,508	4.0	917,702	14.6	1,014,410	13.4	912,205	13.7
Community	156	0.9	7,612,464	7.5	1,168,849	18.6	711,094	9.4	711,586	10.7
Operating	1,143	6.6	5,779,887	5.7	302,208	4.8	500,106	6.6	155,021	2.3
Total	**17,430**	**100.0**	**$101,137,232**	**100.0**	**$6,300,512**	**100.0**	**$7,557,336**	**100.0**	**$6,679,543**	**100.0**

WEST

Foundation Type	No. of Foundations	%	Assets	%	Gifts Received	%	Qualifying Distributions[1]	%	Total Giving[2]	%
Independent	10,145	87.5	$104,853,010	81.2	$5,952,633	80.4	$5,852,972	75.7	$5,425,760	81.8
Corporate	353	3.0	1,671,971	1.3	344,997	4.7	334,909	4.3	320,289	4.8
Community	110	0.9	7,348,956	5.7	860,045	11.6	604,404	7.8	601,483	9.1
Operating	990	8.5	15,209,653	11.8	244,425	3.3	934,575	12.1	288,141	4.3
Total	**11,598**	**100.0**	**$129,083,591**	**100.0**	**$7,402,100**	**100.0**	**$7,726,859**	**100.0**	**$6,635,673**	**100.0**

Source: The Foundation Center, *Foundation Yearbook,* 2005. Geographic regions as defined by U.S. Census Bureau. Table excludes eleven foundations based in the Caribbean and two based in the South Pacific. Due to rounding, figures may not add up. Dollars in thousands.

[1]Qualifying distributions are the expenditures used in calculating the required 5 percent payout for private foundations; includes total giving, as well as reasonable administrative expenses, set-asides, PRIs, operating program expenses, and the amount paid to acquire assets used directly for charitable purposes.

[2]Includes grants, scholarships, and employee matching gifts; excludes set-asides, loans, PRIs, and program expenses. For some operating foundations, program expenses are included.

Within the Midwest region, close to seven out of ten (68.3 percent) corporate funders were located in the East North Central subregion (IL, IN, MI, OH, and WI) and they provided a similar share (69.6 percent) of corporate giving in the Midwest region. In the Northeast, corporate foundations in the Middle Atlantic subregion (NJ, NY, and PA) accounted for close to seven out of ten Northeastern corporate funders (67.4 percent) and more than four out of five corporate grant dollars (82.8 percent). Similarly, corporate foundations in the relatively large South Atlantic subregion (DE, DC, FL, GA, MD, NC, SC, VA, and WV) dominated in the South with three-fifth (59.9 percent) of the share of foundations in the region. In the West, corporate foundations in the Pacific subregion provided a much larger share (72 percent) of support. However, giving by corporate foundations in only one state—California—represented close to two-thirds (64 percent) of all corporate foundation giving in the West.

While the Midwest held the largest number of corporate foundations by region, at the state level, New York led in number of corporate foundations (253). Moreover, these foundations provided close to one-in-six U.S. corporate foundation grant dollars (16.4 percent)—more than the combined share of New Jersey and Illinois, the second- and third-ranked states by corporate foundation giving. Following these states by share of all corporate foundation giving were Michigan (6.2 percent) and Texas (6 percent).

Finally, several states that did not rank near the top in terms of their share of all U.S. corporate foundation resources nonetheless counted on these funders for a relatively large share of their state's foundation support. The top five states based on corporate foundation giving as a share of all 2003 foundation giving in the state included Arkansas (37.6 percent), Delaware (28.5 percent), Minnesota (25.7 percent), North Carolina (23.3 percent), and Oregon (22.4 percent).

Community Foundations

- **Close to half of U.S. community foundations were based in Midwest in 2003**

- **Community foundations in the Midwest and South provided largest shares of community foundation giving by region**

- **California ranked first by share of overall U.S. community foundation giving**

Community foundations are far more concentrated in the Midwest than in the other major regions. Close to half (48.1 percent) of the 699 grantmaking U.S. community foundations were based in the 12-state region (Table 13). In addition, Figure 25 shows that Midwestern community funds accounted for a larger share of overall giving within their region (12.3 percent). Still, community foundations in the South provided 10.7 percent of giving in that region, approaching the Midwest's share. By comparison, community foundations in the West awarded 9.1 percent of grant dollars, while Northeastern community foundations accounted for a much more modest 4 percent of giving.

More than three-quarters (77.1 percent) of Midwestern community foundations were located in the East North Central subregion (IL, IN, MI, OH, and WI) in 2003, and they provided over two-thirds (68.3 percent) of the region's community foundation giving. In the Northeast, community foundations in the Middle Atlantic subregion (NJ, NY, and PA) accounted for over three-fifths (62.1 percent) of the number of funders in the region and a smaller share (57.2 percent) of Northeast community foundation giving. Led by the New York Community Trust—the nation's largest community foundation—New York community foundations alone provided more than one-third (36.2 percent) of all community foundation giving in the Northeast

FIGURE 25.   Distribution of Foundation Giving by Type and Region, 2003

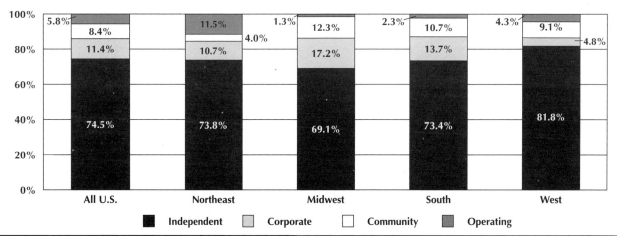

Source: The Foundation Center, *Foundation Yearbook*, 2005. Due to rounding, percentages may not total 100.

TABLE 13.   Fiscal Data of Corporate and Community Grantmaking Foundations by Region and State, 2003

Region[1]	Corporate Foundations						Community Foundations					
	No.	%	Assets	%	Total Giving[2]	%	No.	%	Assets	%	Total Giving[2]	%
NORTHEAST	733	28.8	$ 4,946,682	32.0	$1,126,498	32.5	95	13.6	$ 6,568,256	19.2	$ 424,919	16.8
New England	239	9.4	1,031,889	6.7	193,725	5.6	36	5.2	2,842,968	8.3	181,804	7.2
Connecticut	61	2.4	174,555	1.1	81,209	2.3	18	2.6	1,012,464	3.0	51,416	2.0
Maine	11	0.4	45,452	0.3	6,114	0.2	2	0.3	135,652	0.4	10,055	0.4
Massachusetts	122	4.8	660,650	4.3	82,005	2.4	12	1.7	917,587	2.7	72,315	2.9
New Hampshire	15	0.6	15,847	0.1	2,364	0.1	2	0.3	288,298	0.8	17,044	0.7
Rhode Island	25	1.0	128,289	0.8	19,731	0.6	1	0.1	393,510	1.2	24,974	1.0
Vermont	5	0.2	7,096	0.0	2,301	0.1	1	0.1	95,457	0.3	6,000	0.2
Middle Atlantic	494	19.4	3,914,793	25.3	932,774	26.9	59	8.4	3,725,288	10.9	243,115	9.6
New Jersey	87	3.4	607,559	3.9	259,700	7.5	5	0.7	162,501	0.5	15,942	0.6
New York	253	9.9	2,390,242	15.5	567,998	16.4	19	2.7	2,318,810	6.8	153,754	6.1
Pennsylvania	154	6.0	916,992	5.9	105,076	3.0	35	5.0	1,243,977	3.6	73,419	2.9
MIDWEST	877	34.4	4,787,699	31.0	1,106,613	31.9	336	48.1	12,603,728	36.9	791,538	31.3
East North Central	599	23.5	3,369,583	21.8	770,172	22.2	259	37.1	9,350,222	27.4	540,271	21.3
Illinois	176	6.9	989,017	6.4	230,250	6.6	20	2.9	1,460,318	4.3	64,891	2.6
Indiana	57	2.2	254,615	1.6	54,819	1.6	74	10.6	1,560,043	4.6	63,074	2.5
Michigan	84	3.3	732,816	4.7	215,091	6.2	65	9.3	1,695,803	5.0	92,041	3.6
Ohio	148	5.8	902,191	5.8	181,735	5.2	69	9.9	3,851,644	11.3	253,658	10.0
Wisconsin	134	5.3	490,945	3.2	88,277	2.5	31	4.4	782,414	2.3	66,608	2.6
West North Central	278	10.9	1,418,116	9.2	336,441	9.7	77	11.0	3,253,506	9.5	251,267	9.9
Iowa	66	2.6	291,579	1.9	29,065	0.8	18	2.6	154,635	0.5	14,614	0.6
Kansas	27	1.1	26,593	0.2	9,576	0.3	14	2.0	118,460	0.3	8,738	0.3
Minnesota	90	3.5	668,157	4.3	172,989	5.0	13	1.9	1,416,748	4.1	89,378	3.5
Missouri	68	2.7	377,657	2.4	101,135	2.9	8	1.1	981,426	2.9	83,708	3.3
Nebraska	20	0.8	46,609	0.3	22,698	0.7	16	2.3	430,945	1.3	43,437	1.7
North Dakota	4	0.2	3,073	0.0	771	0.0	4	0.6	62,903	0.2	2,482	0.1
South Dakota	3	0.1	4,447	0.0	207	0.0	4	0.6	88,387	0.3	8,910	0.4
SOUTH	586	23.0	4,040,508	26.2	912,205	26.3	156	22.3	7,612,464	22.3	711,586	28.1
South Atlantic	351	13.8	2,308,487	14.9	510,918	14.7	94	13.4	3,690,025	10.8	348,491	13.8
Delaware	6	0.2	106,216	0.7	55,751	1.6	1	0.1	124,579	0.4	5,723	0.2
District of Columbia	12	0.5	339,959	2.2	49,208	1.4	1	0.1	311,715	0.9	76,463	3.0
Florida	57	2.2	289,675	1.9	37,297	1.1	24	3.4	815,356	2.4	50,815	2.0
Georgia	53	2.1	388,017	2.5	103,606	3.0	11	1.6	452,053	1.3	52,373	2.1
Maryland	40	1.6	156,221	1.0	28,978	0.8	7	1.0	177,708	0.5	22,916	0.9
North Carolina	88	3.5	425,943	2.8	160,807	4.6	19	2.7	945,040	2.8	70,605	2.8
South Carolina	25	1.0	93,956	0.6	11,178	0.3	6	0.9	270,036	0.8	20,390	0.8
Virginia	58	2.3	488,900	3.2	62,209	1.8	15	2.1	425,830	1.2	43,092	1.7
West Virginia	12	0.5	19,601	0.1	1,885	0.1	10	1.4	167,707	0.5	6,114	0.2
East South Central	85	3.3	745,510	4.8	71,244	2.1	25	3.6	1,011,610	3.0	171,785	6.8
Alabama	29	1.1	223,907	1.4	27,440	0.8	10	1.4	183,454	0.5	37,629	1.5
Kentucky	20	0.8	79,227	0.5	8,702	0.3	5	0.7	203,887	0.6	10,816	0.4
Mississippi	11	0.4	34,195	0.2	2,508	0.1	5	0.7	33,824	0.1	3,851	0.2
Tennessee	25	1.0	408,182	2.6	32,594	0.9	5	0.7	590,445	1.7	119,489	4.7
West South Central	150	5.9	986,511	6.4	330,043	9.5	37	5.3	2,910,828	8.5	191,310	7.6
Arkansas	14	0.5	52,210	0.3	102,417	3.0	3	0.4	67,096	0.2	3,348	0.1
Louisiana	17	0.7	23,718	0.2	9,066	0.3	3	0.4	282,209	0.8	16,951	0.7
Oklahoma	16	0.6	70,306	0.5	9,350	0.3	5	0.7	1,257,919	3.7	33,382	1.3
Texas	103	4.0	840,277	5.4	209,211	6.0	26	3.7	1,303,604	3.8	137,628	5.4
WEST	353	13.8	1,671,971	10.8	320,289	9.2	110	15.7	7,348,956	21.5	601,483	23.8
Mountain	99	3.9	257,674	1.7	28,752	0.8	32	4.6	1,328,839	3.9	92,236	3.6
Arizona	16	0.6	54,462	0.4	3,989	0.1	3	0.4	423,615	1.2	24,455	1.0
Colorado	24	0.9	40,439	0.3	12,104	0.3	15	2.1	624,233	1.8	36,644	1.4
Idaho	9	0.4	102,675	0.7	4,668	0.1	1	0.1	48,614	0.1	3,523	0.1
Montana	6	0.2	14,740	0.1	2,852	0.1	2	0.3	45,298	0.1	933	0.0
Nevada	11	0.4	6,148	0.0	1,843	0.1	3	0.4	40,232	0.1	6,337	0.3
New Mexico	16	0.6	17,060	0.1	857	0.0	5	0.7	70,838	0.2	4,967	0.2
Utah	16	0.6	17,343	0.1	1,932	0.1	1	0.1	764	0.0	582	0.0
Wyoming	1	0.0	4,807	0.0	508	0.0	2	0.3	75,245	0.2	14,795	0.6
Pacific	254	10.0	1,414,298	9.2	291,537	8.4	78	11.2	6,020,117	17.6	509,248	20.1
Alaska	17	0.7	56,640	0.4	2,785	0.1	3	0.4	14,471	0.0	2,074	0.1
California	166	6.5	1,020,958	6.6	205,065	5.9	49	7.0	4,687,408	13.7	412,130	16.3
Hawaii	16	0.6	46,986	0.3	8,541	0.2	2	0.3	151,163	0.4	16,088	0.6
Oregon	26	1.0	184,671	1.2	48,194	1.4	8	1.1	657,227	1.9	34,693	1.4
Washington	29	1.1	105,043	0.7	26,952	0.8	16	2.3	509,848	1.5	44,263	1.7
CARIBBEAN[3]	0	0.0	0	0.0	0	0.0	2	0.3	19,576	0.1	2,003	0.1
Puerto Rico	0	0.0	0	0.0	0	0.0	1	0.1	16,990	0.0	712	0.0
Virgin Islands	0	0.0	0	0.0	0	0.0	1	0.1	2,587	0.0	1,291	0.1
SOUTH PACIFIC[3]	0	0.0	0	0.0	0	0.0	0	0.0	0	0.0	0	0.0
American Samoa	0	0.0	0	0.0	0	0.0	0	0.0	0	0.0	0	0.0
Total	2,549	100.0	$15,446,861	100.0	$3,465,605	100.0	699	100.0	$34,152,980	100.0	$2,531,530	100.0

Source: The Foundation Center, *Foundation Yearbook,* 2005. Due to rounding, figures may not add up. Dollars in thousands.
[1]Geographic regions are those defined by the U.S. Census Bureau.
[2]Includes grants, scholarships, and employee matching gifts; excludes set-asides, loans, PRIs, and program expenses.
[3]Private foundations in Puerto Rico, the Virgin Islands and American Samoa are not required to file Form 990-PF. Only a few voluntary reporters are represented.

region. The South Atlantic subregion (DE, DC, FL, GA, MD, NC, SC, VA, and WV) held the largest share of community foundations in the South by number (60.3 percent) and close to half the giving (49 percent). Finally, California, located in the Pacific subregion of the West, alone accounted for well over two-fifths (44.5 percent) of the number of Western community foundations and more than two-thirds (68.2 percent) of their giving in the region.

The Midwest held the largest number of community foundations by region and was also home to the three largest states by number of U.S. community foundations: Indiana (74), Ohio (69), and Michigan (65). California ranked fourth by number (49), but first by share of all community foundation giving. Overall, California-based community funds gave more than one-in-six U.S. community foundation grant dollars in 2003 (16.3 percent). Ohio ranked second, accounting for one-in-ten community foundation grant dollars (10 percent), followed by New York (6.1 percent), Texas (5.4 percent), and Tennessee (4.7 percent).

Several states counted on community foundations for a relatively large share of their state's foundation funding, although they did not rank near the top in terms of their share of all U.S. community foundation resources. The top five states based on community foundation giving as a share of all 2003 foundation giving in the state included: South Dakota (44.4 percent), Tennessee (37.1 percent), North Dakota (34.4 percent), New Hampshire (34.1 percent), and Vermont (30.3 percent).

New Foundation Creation

The development of new foundations exerts a profound impact on asset growth patterns, and thus on the grantmaking potential of the foundation community. Individuals who establish foundations during their lifetimes may opt to transfer into their foundations only the amount they choose to pay out in grants each year, much as any individual making a private contribution. Alternatively, they may transfer a share of their personal wealth into a permanent endowment for future charitable use, thereby committing the foundation to a more predictable level of spending. In many cases, foundations that begin as the former— also known as "pass-through" foundations—will become fully endowed later in their existence, often as a result of a donor's bequest. Figure 26 presents the establishment of mid-size and larger foundations by decade and shows that, in general, the creation of philanthropies in the West and South lagged behind the Northeast and the Midwest until the 1970s, when the formation of foundations in every region experienced a pronounced decline. Much of this decline can be attributed to more restrictive foundation regulations and to a stagnant economy. New formation rebounded in the 1980s, with about 19 percent of larger active U.S. foundations created during the decade, and similar shares reported for the Midwest (20.5 percent), South (18.8 percent), Northeast (18.3 percent), and West (18.1 percent).

Well over two-fifths (44.2 percent) of larger active Western foundations were established in the 1990s,

FIGURE 26. Decade of Establishment for Larger U.S. Foundations by Region

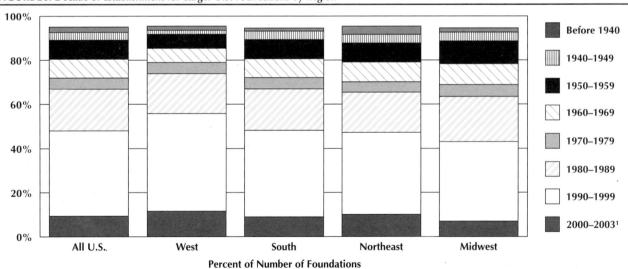

Legend:
- Before 1940
- 1940–1949
- 1950–1959
- 1960–1969
- 1970–1979
- 1980–1989
- 1990–1999
- 2000–2003[1]

Percent of Number of Foundations

Source: The Foundation Center, *Foundation Yearbook,* 2005. Based on Foundation Center survey of 20,827 grantmaking foundations with assets of at least $1 million or making grants of $100,000 or more in 2002-2003. Establishment data was not available for 1,053 foundations, including 171 in the West, 293 in the South, 328 in the Northeast, and 260 in the Midwest.
[1] Data incomplete for 2000–2003.

followed by 39.2 percent of Southern foundations. Close to two-fifths (38.6 percent) of all U.S. foundations were established in the 1990s, with slightly smaller shares reported for Northeastern (37.1 percent) and Midwestern (36 percent) grantmakers. To a significant degree, the surge in foundation formation in the 1990s reflected the phenomenal explosion of wealth in the United States during this period.

The pace of foundation establishment slowed between 2000 and 2003, mirroring the economic decline of the period. In 2003, the number of grantmaking foundations grew by 1,555, or 2.4 percent—the smallest increase in number since 1996 and the lowest rate of growth since 1990. If new formation continues at this rate, the foundation universe is unlikely to expand to the extent seen in the 1980s or 1990s. Still, 9.4 percent of currently active larger foundations have been formed since 1999. This percentage is greater than the share of total number of larger foundations reported in any decade prior to the 1980s.

CHAPTER 4

Foundations by Foundation Type

The use of the term "foundation" can sometimes prove confusing to the public and even to grantmakers themselves. Not only are there important operational and legal distinctions between private foundations and other types of charitable organizations, but sharp contrasts exist between different types of private foundations (Figure 27). Foundation type, as much as foundation size, strongly influences patterns of giving and growth.

In general, the Foundation Center defines a private foundation as a nonprofit, nongovernmental organization with a principal endowment of its own that maintains or aids charitable, educational, religious, or other activities serving the public good, primarily by making grants to other nonprofit organizations. Private foundations derive their funds from a single source, such as an individual, a family, or a corporation, and their activities are governed by specific

FIGURE 27. General Characteristics of Four Types of Foundations

Foundation Type	Description	Source of Funds	Decision-Making Activity	Grantmaking Parameters	Reporting
Independent	An independent grant-making organization established to aid social, educational, religious, or other charitable activities.	Endowment generally derived from a single source such as an individual, a family, or a group of individuals. Contributions to endowment limited as to tax deductibility.	Decisions may be made by donor or members of the donor's family; by an independent board of directors or trustees; or by a bank or trust officer acting on the donor's behalf.	Broad discretionary giving allowed but may have specific guidelines and give only in a few specific fields. About three out of four limit their giving to local area.	Annual information returns (990-PF) filed with IRS must be made available to public. A small percentage issue separately printed annual reports.
Corporate	Legally an independent grantmaking organization with close ties to the corporation providing funds.	Endowment and annual contributions from a profit-making corporation. May maintain small endowment and pay out most of contributions received annually in grants, or may maintain endowment to cover contributions in years when corporate profits are down.	Decisions made by board of directors often composed of corporate officials, but which may include individuals with no corporate affiliation. Decisions may also be made by local company officials.	Giving tends to be in fields related to corporate activities or in communities where corporation operates. Usually give more grants but in smaller dollar amounts than independents with comparable giving.	Same as above.
Operating	An organization that uses its resources to conduct research or provide a direct service.	Endowment usually provided from a single source, but eligible for maximum deductible contributions from public.	Decisions generally made by independent board of directors.	Makes few, if any, grants. Grants generally related directly to the foundation's program.	Same as above.
Community	A publicly sponsored organization that makes grants for social, educational, religious, or other charitable purposes in a specific community or region.	Contributions received from many donors. Usually eligible for maximum tax deductible contributions from public.	Decisions made by board of directors representing the diversity of the community.	Grants generally limited to charitable organizations in local community.	Annual information returns (990) filed with the IRS must be made available to public. Many publish full guidelines or annual reports.

Source: The Foundation Center, *Foundation Yearbook*, 2005.

rules and regulations set in the tax code (see Appendix B, "Regulation of Private Foundations").[1]

Some private foundations are organized as operating foundations, which conduct their own research programs or provide a direct service. Although it is not their primary function, many operating foundations occasionally make grants to other nonprofit organizations. In general, however, only a few maintain grant programs of any size.

Community foundations function in much the same way as private foundations, but because their funds are drawn from many donors, they are usually classified as public charities.

The Center's analysis of trends since 1975 has been primarily concerned with three types of foundations: independent, corporate, and community. The emergence in the 1990s of several large grantmaking operating foundations prompted the expansion of this analysis to include these foundations, beginning with 1995 data.

Trends by Foundation Type Through 2003

The following analyses examine changes in foundation resources by foundation type—independent, corporate, community, and grantmaking operating—through 2003, the latest year for which complete fiscal information was available. They include comparisons by number of foundations (Figure 28 and Table 14), total giving (Figure 29 and Table 15), assets (Figure 30 and Table 16), and gifts received (Figure 31 and Table 17).[2]

Independent Foundations

Independent foundations comprise the largest and perhaps most diverse segment of the foundation universe and include family-sponsored and "new health" foundations.

TABLE 14. Total Number of Foundations by Type, 2002 to 2003

Foundation Type	2002	2003	% Change
Independent	57,834	58,991	2.0
Corporate	2,362	2,549	7.9
Community	661	699	5.7
Operating	3,986	4,159	4.3
Total	64,843	66,398	2.4

Source: The Foundation Center, *Foundation Yearbook,* 2005.

TABLE 15. Total Giving by Foundation Type, 2002 to 2003

Foundation Type	2002	2003	% Change
Independent	$23,253.7	$22,567.7	(2.9)
Corporate	3,457.2	3,466.6	0.2
Community	2,526.3	2,531.5	0.2
Operating	1,194.7	1,744.0	46.0
Total	$30,431.8	$30,308.8	(0.4)

Source: The Foundation Center, *Foundation Yearbook,* 2005. Total giving includes grants, scholarships, and employee matching gifts; excludes set-asides, loans, PRIs, and program expenses. Dollars in millions. Due to rounding, figures may not add up. Percent change represents current dollars.

FIGURE 28. Number of Foundations by Foundation Type, 2003

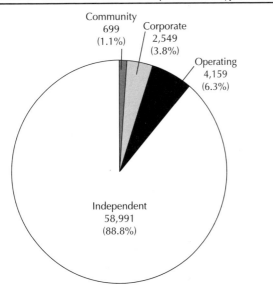

Community 699 (1.1%)
Corporate 2,549 (3.8%)
Operating 4,159 (6.3%)
Independent 58,991 (88.8%)

Total Number = 66,398

Source: The Foundation Center, *Foundation Yearbook,* 2005.

FIGURE 29. Giving by Foundation Type, 2003

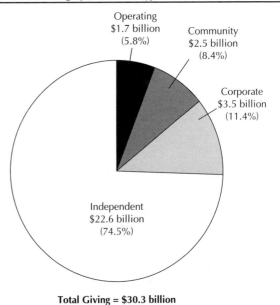

Operating $1.7 billion (5.8%)
Community $2.5 billion (8.4%)
Corporate $3.5 billion (11.4%)
Independent $22.6 billion (74.5%)

Total Giving = $30.3 billion

Source: The Foundation Center, *Foundation Yearbook,* 2005.

Because most of the top-ranking foundations by assets and giving are independent or family-sponsored, trends described in Chapter 2 relating to large foundations apply equally to independents.

2003 Giving

- Giving by independent foundations decreased 2.9 percent

- Independent foundations accounted for nearly three-quarters of grant dollars

- Giving by the top 50 independent foundations remained nearly unchanged

Change in Number of Independent Foundations. A total of 58,991 independent foundations reported grant activity in 2003, up 2 percent from the previous year (Figure 28 and Table 14). The actual number of

independent foundations grew by 1,157—well below the peak increase of 5,708 recorded in 2000. Nonetheless, the number of independent foundations has more than doubled since 1990.

Independent Foundations as a Share of All Grant Dollars. Independent foundations accounted for almost 90 percent of the total number of grantmaking foundations (Figure 28) and were responsible for nearly three-quarters (74.5 percent) of the foundation grant dollars awarded in 2003 (Figure 29). In actual dollars, independent foundations paid out $22.6 billion in grants in the latest year, down from $23.3 billion in the prior year. Still, this represented the third highest level of independent foundation giving on record.

Giving by Larger vs. Smaller Independent Foundations. Of the 58,991 active independent foundations, roughly 18,000, or 31 percent, awarded grants totaling $100,000 or more (Table 18). Yet these foundations distributed

TABLE 16. Assets by Foundation Type, 2002 to 2003

Foundation Type	2002	2003	% Change
Independent	$364.1	$399.1	9.6
Corporate	14.4	15.4	7.1
Community	29.8	34.2	14.7
Operating	26.8	28.0	4.2
Total	**$435.2**	**$476.7**	**9.5**

Source: The Foundation Center, *Foundation Yearbook,* 2005. Dollars in billions. Due to rounding, figures may not add up. Percent change represents current dollars.

TABLE 17. Gifts Received by Foundation Type, 2003 to 2003

Foundation Type	2002	2003	% Change
Independent	$13,951.6	$15,845.9	13.6
Corporate	3,001.8	3,234.1	7.7
Community	3,175.1	3,476.3	9.5
Operating	2,034.9	2,302.1	13.1
Total	**$22,163.4**	**$24,858.3**	**12.2**

Source: The Foundation Center, *Foundation Yearbook,* 2005. Dollars in millions. Due to rounding, figures may not add up. Percent change represents current dollars.

FIGURE 30. Assets by Foundation Type, 2003

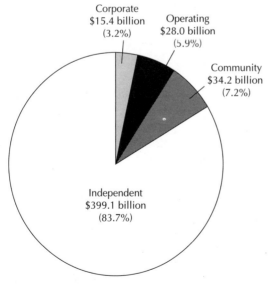

Total Assets = $476.7 billion

Source: The Foundation Center, *Foundation Yearbook,* 2005.

FIGURE 31. Gifts Received by Foundation Type, 2003

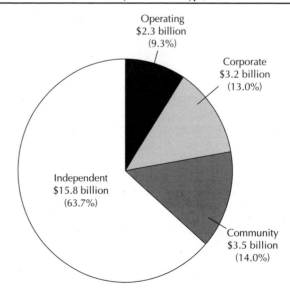

Total Gifts Received = $24.9 billion

Source: The Foundation Center, *Foundation Yearbook,* 2005.

$21.5 billion, or about 95 percent of the grant dollars paid in 2003 by all independent foundations. Within this group of funders, however, the pattern of giving varied considerably. For example, giving by foundations awarding grants totaling $25 million to $100 million dropped 26.7 percent. In contrast, the biggest funders (those awarding grants totaling $100 million or more) posted an 8.1 percent increase in giving, while total giving by smaller funders (those awarding grants totaling $100,000 to $1 million) rose 3.1 percent. By comparison, the remaining 41,000 foundations awarding grants of under $100,000 showed a 0.9 percent dip in giving.

Established vs. New Independent Foundations' Support. The contribution of newly formed independent funders helped mitigate the decrease in overall giving by independent foundations. Numbering nearly 2,000, newly reporting independent foundations added $205.8 million to the overall giving by independent foundations. Excluding this group, giving by established funders fell 3.8 percent.

Funding by the Top 50 Independent Foundations. Support provided by the nation's 50 largest independent foundations by giving remained nearly unchanged at $7.1 billion in 2003. The pattern of giving, however, varied widely among the top funders. Twenty-three raised their levels of giving, while 27 lowered them. The biggest percentage increase was posted by the Moody Foundation. The Texas-based foundation reported an over fivefold jump in its giving, from $9.3 million to $49 million. Much of this growth was due to a $46.6 million grant to Moody Gardens to expand its convention center. Another foundation that showed exceptionally strong growth in 2003 support was the F.W. Olin Foundation. The Florida-based foundation's giving more than doubled from $54.9 million in 2002 to $110.6 million in the latest year. Most of this support came through a $106.6 million grant to the Franklin W. Olin College of Engineering for general and capital

support.[3] Fourteen other foundations among the top 50 foundations by total giving experienced double-digit growth in their 2003 giving.

Among the 27 independent foundations that decreased their giving, the largest decline was reported by the James Irvine Foundation. The California-based foundation's giving dropped 24.4 percent, from $73.5 million in 2002 to $55.5 million in 2003. By comparison, its giving grew by roughly one-quarter between 2001 and 2002. Two other foundations experienced a decline of at least 20 percent in their giving, including the Skirball Foundation (NY), whose giving fell 22.1 percent to $69.4 million, and the David and Lucile Packard Foundation (CA), whose giving decreased 20.6 percent to $277.9 million.

To qualify for ranking among the top 50 independent funders, the minimum spending level was $33 million, reported by the Ellison Medical Foundation (MD), compared to the maximum of $1.2 billion, distributed by the Gates Foundation. In addition to Gates, seven other funders gave at least $200 million in 2003, while fifteen paid out at least $100 million. Among the top ten funders, the threshold for giving rose to $195 million from last year's $192 million.

Independent Foundations New to the Top 50. Eight foundations were new to the top 50 independents by total giving in 2003, compared to ten that were new in 2002. These included the 38th-ranked Wallace Foundation (NY), with $49.3 million in total giving; 39th-ranked Moody Foundation, with $49 million in total giving; 42nd-ranked Ave Maria Foundation (MI), with $41.9 million in total giving; 45th-ranked Brown Foundation (TX), with $38.5 million in total giving; 47th-ranked J.E. and L.E. Mabee Foundation (OK), with $35.7 million in total giving; 48th-ranked Weingart Foundation (CA), with $34.2 million in total giving; 49th-ranked Henry Luce Foundation (NY), with $34 million; and 50th-ranked Ellison Medical Foundation, with $33 million in total giving.

2003 Assets

- Independent foundation assets rose 9.6 percent

- 40 independent foundations reported assets of at least $1 billion

- Moore and Dell foundations experienced largest increases in assets

Change in Independent Foundation Assets. The 58,991 active independent foundations held assets totaling $399.1 billion in 2003, a 9.6 percent rise from 2002 (Table 16). This rebound in assets followed a 7.6 percent decrease in the prior year and reflected the stock

TABLE 18. Analysis of Independent Foundations by Total Giving Range, 2003

Total Giving Range	Number of Foundations	%	Total Giving[1]	%
$100 million+	24	0.0	$ 5,680,173	25.2
$25 million–$100 million	53	0.1	2,264,224	10.0
$10 million–$25 million	166	0.3	2,570,544	11.4
$1 million–$10 million	2,549	4.3	6,590,182	29.2
$100,000–$1 million	15,352	26.0	4,377,681	19.4
Under $100,000	40,846	69.2	1,084,947	4.8
Total	**58,991**	**100.0**	**$22,567,751**	**100.0**

Source: The Foundation Center, *Foundation Yearbook,* 2005. Dollars in thousands.
[1]Includes grants, scholarships, and employee matching gifts; excludes set-asides, loans, PRIs, and program expenses.

market recovery and improved economic climate. Still, independent foundation assets remained below the $408.7 billion peak level recorded in 2000.

Assets of Large vs. Small Independent Foundations. About 35 percent of independent foundations (20,529) reported assets of at least $1 million, yet collectively they held 97.3 percent of the assets of independent foundations overall (Table 19). Only about 6.6 percent of independent foundations (3,907) maintained assets of $10 million or more, but together they accounted for 84.4 percent of independent foundation assets and 73.2 percent of independent foundation giving. In contrast, foundations with assets of under $1 million represented 65.1 percent of independent foundations, but they held less than 3 percent of independent foundation assets and accounted for 9.2 percent of independent foundation giving. Much of the growth of independent foundation assets in the latest year was concentrated among the largest foundations. For example, assets of independent foundations holding assets of $1 billion or more jumped 19.5 percent in 2003, while assets of foundations with assets of under $1 million rose a modest 1.2 percent.

Assets of the Top 50 Independent Foundations. Assets of the 50 largest independent foundations rose 17.3 percent in 2003, with the median increase being 14.6 percent. By comparison, total assets of all independent foundations rose 9.6 percent. Six grantmakers—the Gordon and Betty Moore (CA), Bill & Melinda Gates (WA), David and Lucile Packard (CA), William and Flora Hewlett (CA), Andrew W. Mellon (NY), and W. K. Kellogg (MI) foundations—accounted for over half of the increase in the assets of the largest 50 foundations. Excluding these six foundations, assets of the remaining top funders grew by 11.5 percent.

Among the independent foundations in the top 50 list, forty-seven realized asset gains, up from only nine in the prior year. The ninth-ranked Moore Foundation posted the largest percentage increase, with assets ballooning more than fiftyfold, from $93.3 million to $4.8 billion. The assets of the California-based foundation were boosted by a $3.7 billion gift from its founders—the largest amount of gifts into a foundation in

2003. The second highest rate of asset growth was reported by the 40th-ranked Michael and Susan Dell Foundation, whose assets more than doubled from $463.9 million to $1.1 billion. The Texas-based foundation benefited from a $637.7 million gift from its founders.

Among the largest 50 foundations by size of endowment, three reported asset losses in 2003. The biggest drop was recorded by the Donald W. Reynolds Foundation (NV), whose assets declined 10.3 percent to $1.2 billion. The other funders reporting decreased assets were the Moody Foundation (TX), with assets down by 3.7 percent to $931.7 million, and the Robert Wood Johnson Foundation (NJ), with assets slipping by 1 percent to $7.9 billion.

Qualifying asset values among the 50 largest independents ranged from the Gates Foundation's $26.8 billion to the Weingart Foundation's (CA) $749 million. By comparison, the minimum asset value for the top independent foundations in 2002 was $670.9 million. In addition, 40 independent foundations held assets of at least $1 billion, up from 35 in the prior year.

Independent Foundations New to the Top 50. Four foundations were new to the list of top 50 independents by asset size in 2003. Aside from the Moore and Dell foundations, whose assets realized exceptional growth, the new foundations included the 48th-ranked Conrad N. Hilton Foundation (NV), whose assets grew 29.1 percent to $764 million, and 49th-ranked Joyce Foundation (IL), whose assets rose 15.2 percent to $753.1 million.

2003 Gifts Received

• **202 independent foundations received gifts of $10 million or more, down from 216 in 2002**

• **Moore Foundation received largest gift in 2003**

Gifts and Bequests to Independent Foundations. Bequests to independent foundations, combined with gifts from living donors, rose 13.6 percent to $15.8 billion in 2003 (Table 17). This gain followed a 32.1 percent decline in the prior year. Still, the level of gifts received by independent foundations in the latest year

TABLE 19. Analysis of Independent Foundations by Asset Range, 2003

Asset Range	Number of Foundations	%	Assets	%	Total Giving[1]	%
$1 billion+	40	0.1	$146,507,919	36.7	$ 6,356,662	28.2
$250 million–$1 billion	113	0.2	52,257,798	13.1	2,316,547	10.3
$50 million–$250 million	725	1.2	74,081,446	18.6	3,981,146	17.6
$10 million–$50 million	3,029	5.1	64,163,721	16.1	3,875,195	17.2
$1 million–$10 million	16,622	28.2	51,343,610	12.9	3,955,195	17.5
Under $1 million	38,462	65.1	10,783,941	2.6	2,083,007	9.2
Total	**58,991**	**100.0**	**$399,138,435**	**100.0**	**$22,567,751**	**100.0**

Source: The Foundation Center, *Foundation Yearbook*, 2005. Dollars in thousands. Due to rounding, figures may not add up.
[1]Includes grants, scholarships, and employee matching gifts; excludes set-asides, loans, PRIs, and program expenses.

remained far below the record of $24.1 billion reported in 1999.

Seven independent foundations received gifts totaling at least $100 million, down slightly from eight in 2002. The overall number of independent foundations benefiting from gifts of $10 million or more decreased to 202 in 2003, from 216 in 2002. Nonetheless, the value of these gifts grew from $7.3 billion to $9.4 billion, reflecting a stark difference in the value of the largest gifts between the two years. In the latest year, the largest gift into a foundation was $3.7 billion, compared to $350 million in 2002 (see below).

Among the 202 independent foundations receiving gifts totaling at least $10 million in 2003, 22 were based in California and in New York, followed by Texas (14), Ohio (eight), Illinois (seven), and New Jersey (six). None of the remaining states reported more than five foundations with gifts in this range.

Largest Gifts Received. The Gordon and Betty Moore Foundation (CA)—which funds primarily in the areas of environmental conservation, science, and higher education—received the largest gift among grantmaking foundations in 2003. The $3.7 billion infusion from Intel co-founder, Gordon E. Moore, and his wife, Betty, catapulted the foundation's assets to $4.8 billion, from $93.3 million in 2002. The second largest gift was reported by the Michael and Susan Dell Foundation, which received $673.7 million from its founders, resulting in a more than doubling of its assets from $463.9 million to $1.1 billion.

Corporate Foundations

Like independent foundations, corporate foundations are classified as private foundations and are therefore subject to the same regulatory and reporting requirements applicable to that tax status (see Appendix B). Their basic operations, however, differ substantially. A corporate foundation generally maintains close ties with the parent company that provides its endowment and receives annual or periodic contributions from that corporation based on yearly profits. A corporate foundation's board and staff are usually composed of individuals who also hold positions within the parent company, and its giving program often reflects corporate interests and is targeted to geographic areas in which the parent company operates.

Foundations represent only one of many channels corporations use to make charitable contributions. Corporations may also make both cash and in-kind contributions—e.g., equipment, office space, staff time—through a direct giving program administered wholly within the company and funded from the company's pre-tax earnings. They may also provide support for charitable organizations through their corporate marketing budgets.

The following statistics cover only the 2,549 corporate foundations reporting complete data to the IRS. Key information on these foundations and over 1,200 direct giving programs is available, however, in three Foundation Center resources: the National Directory of Corporate Giving, FC Search, *and* The Foundation Directory Online.

2003 Giving

- **Corporate foundation giving inched up to a record $3.47 billion**

- **Giving by the top 50 corporate funders rose 4 percent**

- **Corporate foundations accounted for 11 percent of grant dollars**

Change in Number of Corporate Foundations. Between 2002 and 2003, the number of grantmaking corporate foundations increased from 2,362 to 2,549. This 7.9 percent gain surpassed the rates of growth of community (5.7 percent), operating (4.3 percent), and independent (2 percent) foundations (Table 14). Since 1990, the number of corporate foundations has grown by nearly 50 percent.

Corporate Foundations as a Share of All Giving. Corporate foundations accounted for only 3.8 percent of the total number of U.S. grantmaking foundations in 2003 (Figure 28) but were responsible for 11.4 percent of grant dollars awarded (Figure 29) and 13 percent of gifts received by foundations (Figure 31). However, corporate foundations represent a diminishing share of foundation philanthropy. In 1988, for example, they accounted for 17 percent of total giving. Nonetheless, the $3.47 billion in giving in 2003 represented the largest amount of giving recorded for corporate foundations.

Established vs. New Corporate Foundations' Support. Newly established funders helped to shore up corporate giving in 2003. Just over 100 newly reporting corporate foundations added $32.1 million to overall corporate foundation giving. Excluding the grants of this group of corporate foundations, giving by established funders slipped 0.7 percent.

Giving by Large vs. Small Corporate Foundations. Of the 2,549 active corporate foundations, 53.9 percent awarded grants totaling $100,000 or more in 2003 (Table 20). This group was responsible for almost all (98.8 percent) of the grant dollars paid in 2003 by corporate foundations.

Although overall corporate giving inched up 0.2 percent in the latest year, the pattern of giving was highly uneven among funders by size of total giving. The fastest growth in giving was posted by foundations making grants totaling under $100,000 (up 7.4 percent), followed by those awarding grants totaling $100,000 to under $1 million (up 6.7 percent). In contrast, giving by corporate funders awarding grants totaling $10 million to under $25 million—which accounted for about one-fifth of total corporate giving in 2003—decreased by 10.3 percent, while those in the $25 million or more range posted a 4.3 percent rise in total giving.

Funding by the Top 50 Corporate Foundations. Grantmaking by the top 50 corporate foundations by total giving rose 4.2 percent in 2003, compared to a 0.2 percent uptick for corporate foundations overall. There was a wide disparity, however, in the pattern of giving among foundations in the top 50 list. Twenty-seven foundations posted increases in giving, led by the second-ranked Aventis Pharmaceuticals Health Care Foundation (NJ). The foundation's giving nearly tripled from $37.9 million in 2002 to $102.5 million. Most of its support was provided in the form of medication for low-income patients. The second largest growth in giving was reported by the 50th-ranked Thrivent Financial for Lutherans Foundation, whose giving more than doubled from $6.2 million to $14.1 million. The Wisconsin-based corporate foundation supports protestant churches and agencies. Fourteen other corporate foundations reported double-digit rates of growth in giving.

Of the 50 largest corporate foundations by total giving, 23 decreased their giving in 2003. The largest decline in giving was reported by the AT&T Foundation (NY), whose giving dropped 53.9 percent from $39.5 million to $18.2 million. Twelve other corporate foundations registered double-digit declines in their giving.

To qualify for ranking among the top 50 corporate foundations by total giving, the minimum spending level was $14.1 million, reported by the Thrivent Financial for Lutherans Foundation, compared to the maximum of $119.8 million, distributed by the Arkansas-based Wal-Mart Foundation. In addition to Wal-Mart, seven other funders gave at least $50 million in 2003: the Aventis Pharmaceuticals Health Care Foundation (NJ), Ford Motor Company Fund (MI), and the Bank of America (NC), Verizon (NY), Citigroup (NY), J.P. Morgan Chase (NY), and GE (CT) foundations. Among the top ten corporate foundations in 2003, the threshold for giving increased to $48.7 million from the prior year's $47.3 million.

Corporate Foundations New to the Top 50. Six foundations were new to the top 50 corporate foundations by total giving in 2003: the 43rd-ranked Medtronic Foundation (MN), with giving of $16.6 million; 45th-ranked May Department Stores Company Foundation (MO), with giving of $15.5 million; 47th-ranked Duke Energy Foundation (NC), with giving of $14.9 million; 48th-ranked PepsiCo Foundation (NY), with giving of $14.5 million; 49th-ranked Goldman Sachs Foundation (NY), with giving of $14.3 million; and 50th-ranked Thrivent Financial for Lutherans Foundation, with giving of $14.1 million.

Corporate Foundation Support by Industry. A breakdown of 2003 corporate foundation giving by industry showed that banking institutions provided one-fifth (20.1 percent) of corporate foundation grant dollars (Table 21). The foundations of companies in retail and wholesale trade followed, with about one-in-twelve grant dollars (8.1 percent). Overall, close to three-fifths (58 percent) of corporate foundation giving in the latest year came from foundations sponsored by nonmanufacturing enterprises, while the foundations of manufacturing concerns accounted for over two-fifths (42 percent). Among manufacturing concerns, the top industries by share of giving included pharmaceuticals (7.8 percent), transportation equipment (6.3 percent), and industrial and commercial machinery (6.3 percent).

2003 Assets

- Corporate foundation assets increased 7 percent

- Top 50 corporate foundations reported higher-than-average increase

Change in Corporate Foundation Assets. Assets of the 2,549 corporate foundations rose 7.1 percent in 2003, nearly reversing the 7.4 percent decline in the prior year. Overall, assets increased from $14.4 billion to $15.4 billion between 2002 and 2003. The current level of corporate foundation assets, however, fell short of the peak level of $15.9 billion recorded in 2000. Still, assets of corporate foundations have more than doubled since 1994.

TABLE 20. Analysis of Corporate Foundations by Total Giving Range, 2003

Total Giving Range	Number of Foundations	%	Total Giving[1]	%
$25 million+	24	0.9	$1,210,889	34.9
$10 million–$25 million	45	1.8	709,452	20.5
$1 million–$10 million	385	15.1	1,176,328	33.9
$100,000–$1 million	921	36.1	329,652	9.5
Under $100,000	1,174	46.0	39,284	1.2
Total	**2,549**	**100.0**	**$3,465,605**	**100.0**

Source: The Foundation Center, *Foundation Yearbook*, 2005. Dollars in thousands.

[1]Includes grants, scholarships, and employee matching gifts; excludes set-asides, loans, PRIs, and program expenses.

Assets of Large vs. Small Corporate Foundations. Unlike independents, many corporate foundations do not have large endowments but depend instead on their sponsoring companies for continued contributions. Nonetheless, 1,043 corporate foundations (40.9 percent) held assets of at least $1 million, compared to 973 in 2002. Just 307 corporate funders (12 percent) maintained assets of $10 million or more, yet they provided over three-fifths (63.2 percent) of corporate foundation giving (Table 22).

Assets of the Top 50 Corporate Foundations. Assets of the 50 largest corporate foundations rose 18.5 percent in 2003, far exceeding the 7.1 percent overall increase for corporate foundations. Reflecting the stock market upturn and healthier corporate profits, 38 corporate foundations reported asset gains, compared to only fifteen in 2002. The biggest growth was posted by the Verizon Foundation, whose assets more than quadrupled from $80.2 million to $359.9 million. This raised the foundation's rank from sixteenth to third place. The New York-based foundation's assets were bumped up by $299 million in gifts from its company sponsor—the largest amount of gifts received by corporate foundations in 2003. Other corporate foundations that reported exceptional asset growth included BP Foundation (IL), whose assets jumped from $30.2 million to $78.6 million, and the Intel Foundation (OR), whose assets rose from $40 million to $84.4 million. Among the other corporate foundations in the top 50 list,

TABLE 21.   Corporate Foundation Giving by Industry, 2003

Industrial Classification	Number of Foundations	%	Total Giving[1]	%	Average Giving per Foundation
Chemicals	59	2.3	$ 140,608	4.1	$2,383
Computers/Office Equipment	18	0.7	34,042	1.0	1,891
Food and Agriculture	103	4.0	132,406	3.8	1,285
Industrial and Commercial Machinery (Not Computer)	173	6.8	217,737	6.3	1,259
Paper and Like Products	39	1.5	26,520	0.8	680
Petroleum/Gas/Mining	48	1.9	112,401	3.2	2,342
Pharmaceuticals	31	1.2	270,657	7.8	8,731
Primary Metals	47	1.8	31,541	0.9	671
Printing/Publishing/Media	84	3.3	76,145	2.2	906
Textiles and Apparel	83	3.3	30,639	0.9	369
Transportation Equipment	69	2.7	218,465	6.3	3,166
Other Manufacturing	268	10.5	163,350	4.7	610
Total Manufacturing	**1,022**	**40.1**	**$1,454,513**	**42.0**	**$1,423**
Banking	291	11.4	696,073	20.1	2,392
Finance	156	6.1	222,694	6.4	1,428
Insurance	132	5.2	253,319	7.3	1,919
Retail and Wholesale Trade	331	13.0	281,846	8.1	851
Telecommunications	40	1.6	176,316	5.1	4,408
Transportation	37	1.5	71,196	2.1	1,924
Utilities	92	3.6	124,480	3.6	1,353
Other Services	432	16.9	182,519	5.3	422
Total Nonmanufacturing	**1,511**	**59.3**	**$2,008,442**	**58.0**	**$1,329**
Unspecified	16	0.6	2,650	0.1	166
Total	**2,549**	**100.0**	**$3,465,605**	**100.0**	**$1,360**

Source: The Foundation Center, *Foundation Yearbook*, 2005. Categories are based on the Conference Board's classification of corporations using Standard Industrial Classification codes. Dollars in thousands.
[1]Includes grants, scholarships, and employees matching gifts; excludes set-asides, loans, PRIs, and program expenses.

TABLE 22.   Analysis of Corporate Foundations by Asset Range, 2003

Asset Range	Number of Foundations	%	Assets	%	Total Giving[1]	%
$250 million+	6	0.2	$ 1,848,072	12.0	$ 189,057	5.5
$50 million–$250 million	56	2.2	5,393,188	34.9	987,884	28.5
$10 million–$50 million	245	9.6	5,457,224	35.3	1,013,098	29.2
$1 million–$10 million	736	28.9	2,397,554	15.6	751,676	21.7
Under $1 million	1,506	59.1	350,821	2.3	523,889	15.1
Total	**2,549**	**100.0**	**$15,446,861**	**100.0**	**$3,465,605**	**100.0**

Source: The Foundation Center, *Foundation Yearbook*, 2005. Dollars in thousands. Due to rounding, figures may not add up.
[1]Includes grants, scholarships, and employee matching gifts; excludes set-asides, loans, PRIs, and program expenses.

22 reported double-digit gains. By comparison, for the 12 corporate foundations that realized asset losses, only four experienced double-digit asset declines. The largest loss was posted by the Avon Foundation (NY), with assets down 25.3 percent from $103.7 million to $77.5 million. This loss followed a 75.4 percent increase in Avon's assets between 2001 and 2002.

The threshold for qualifying among the top 50 corporate funders by asset size rose from $55 million in 2002 to $69 million in 2003. Qualifying asset values ranged from the Cargill Foundation's (MN) $69.2 million to the Alcoa Foundation's (PA) $453.1 million. Twenty-six corporate foundations held assets of at least $100 million, up from twenty-one in 2002.

Corporate Foundations New to the Top 50. Seven foundations were new to the current list of top 50 corporate foundations by assets, down from thirteen in 2002. Leading these funders was the 12th-ranked Frist Foundation (TN), with assets of $177.7 million. Other foundations new or returning to the list included: 34th-ranked Thrivent Financial for Lutherans Foundation, with assets of $87.5 million; 35th-ranked Intel Foundation, with assets of $84.4 million; 39th-ranked BP Foundation, with assets of $78.6 million; 44th-ranked Pharmacia Foundation (NJ), with assets of $75.3 million; 47th-ranked Simpson PSB Fund (CA), with assets of $71.8 million; and 50th-ranked Cargill Foundation, with assets of $69.2 million.

2003 Gifts Received

- **Gifts from companies to their foundations rose nearly 8 percent**

- **Payout exceeded pay-in by 7 percent**

Gifts to Corporate Foundations. The aggregate value of gifts from companies to their foundations (pay-in) totaled $3.2 billion in 2003, up 7.7 percent from 2002 (Table 17). This increase followed a 1.3 percent dip in gifts received in 2002. Giving (payout) exceeded pay-in by about $232 million (7.2 percent), requiring corporate foundations to draw on their assets.

A total of 67 corporate foundations received gifts of $10 million or more in 2003. Twenty-nine received gifts of at least $25 million, up from 20 in the previous year. Of those receiving $10 million or more, ten were based in New York, six in Illinois, and four each in California, Massachusetts, Michigan, Missouri, and North Carolina. None of the remaining states reported more than three corporate foundations with gifts in this range.

Largest Gifts Received. The largest corporate gift amount reported in 2003 was $299 million given to the New York-based Verizon Foundation from its corporate parent. Other recipients of especially large gifts included the California-based Wells Fargo Foundation ($165.2 million), followed by the Arkansas-based Wal-Mart Foundation ($130.7 million) and the New Jersey-based Aventis Pharmaceuticals Health Care Foundation ($102.5 million).

Corporate Profits, Giving, and Foundations

In 2003, corporate profits experienced the strongest jump seen since 1995. Adjusted for inflation, pre-tax profits of U.S. corporations rose by 12.4 percent. Figures from *Giving USA* for 2003 showed corporate giving as a percent of pre-tax profits was 1.3 percent. In terms of the dollar value of contributions, giving increased to $11.2 billion (3.6 percent) in 2003, following a 7.5 percent decline in the prior year.

Corporate giving as a percent of pre-tax profits reached a peak of 2 percent in 1986, then steadily declined to a low of 1 percent in 1996, and gradually rose to 1.7 percent in 2001. Since then, the share has been declining. Nonetheless, between 1990 and 2003, the growth in corporate charitable contributions outpaced the growth in corporate pre-tax profits.

Using *Giving USA*'s figure of $11.2 billion in total corporate contributions and the Foundation Center's figure of $3.5 billion in actual corporate foundation giving, 2003 foundation giving represented just over three-tenths of total corporate giving. Still, the $11.2 billion total includes not only cash, but also product, property and equipment, and gifts of securities. Foundation giving—consisting mainly of cash—represents a higher proportion of contributions if only cash gifts are measured.

Looking forward, the Foundation Center estimated a 4 percent increase in 2004 corporate foundation giving. As a result, foundation grantmaking as a share of overall corporate giving is expected to rebound.

1. Giving USA Foundation, *Giving USA 2005: The Annual Report on Philanthropy for the Year 2004*, edited by M. Brown, Giving USA Foundation, Indianapolis, IN, 2005.

Community Foundations

Community foundations represent a relatively small but extremely vital and influential component of the foundation universe. Not only were community foundations one of the fastest growing segments of philanthropy in the 1980s, but their leadership role in many communities increased along with stepped-up demands for private initiatives to solve local problems, such as crime, drug abuse, HIV/AIDS, homelessness, and the failure of many public school systems. In the 1990s, the role of community foundations was further heightened by the impact of economic restructuring and government downsizing on local economies. In the new millennium—a period marked by ballooning budget deficits that curtail government support for nonprofits—community foundations continue to seek ways to expand resources by forging partnerships between public and private funding sources.

2003 Giving

- **Giving by community foundations inched up 0.2 percent**

- **Grantmaking by the 25 largest community foundations increased 8 percent**

- **Community funds accounted for over one-in-twelve grant dollars**

Change in Number of Community Foundations. In 2003, the number of grantmaking community foundations climbed 5.7 percent to 699, which followed a 9.8 percent gain in the prior year. Overall, the number of community foundations in the Foundation Center's census has more than doubled since 1990.[4]

Community Foundations as a Share of All Giving. Community foundations accounted for 1.1 percent of the total number of active grantmaking foundations (Figure 28) but were responsible for 8.4 percent of giving (Figure 29) and 14 percent of gifts received in 2003 (Figure 31). Between 2002 and 2003, giving by community foundations rose 0.2 percent to $2.5 billion. This current level of giving was nearly fourteen times greater than the $183 million in giving reported in 1981, when the Foundation Center started tracking data separately for community foundations. The share of overall foundation giving accounted for by community foundations has steadily increased from roughly 5 percent in the early 1980s to its peak of 8.4 percent in 2003.

Giving by Large vs. Small Community Funds. Of the 699 community foundations identified in the latest year, 251 made grants totaling $1 million or more, up from 242 reported in 2002 (Table 23). These 251 foundations together paid out $2.4 billion in grants, or

about 95 percent of community foundation giving. The remaining 448 funders collectively gave away $117 million.

Although overall community foundation giving rose marginally in 2003, changes in giving were highly uneven based on range of total giving. Specifically, funders awarding grants totaling $25 million to $100 million posted the largest increase in giving (7.3 percent), followed by those with grants totaling $1 million to $10 million (up 6 percent). In contrast, giving by community foundations in the $10 million to $25 million range of total giving decreased 21.5 percent, while giving by those in the top tier—$100 million or more—declined 6.7 percent.

Established vs. New Community Foundations' Support. Newly established funders helped stabilize community foundation giving in 2003. In all, 39 newly reporting community funds contributed $18 million of the $2.5 billion in grant dollars reported by community foundations overall. Excluding the grants of these newly reporting community foundations, giving by established funders slipped 0.5 percent.

Funding by the Top 25 Community Foundations. Grantmaking by the nation's top 25 community foundations by total giving rose 8 percent in 2003, compared to the 0.2 percent uptick reported for community foundations overall. The largest increase among the top 25 funders was reported by the Community Foundation of Greater Birmingham (AL), whose giving more than tripled from $8.6 million to $32.2 million. Contributing to this increase was $10.5 million in gifts into the foundation in 2003. Twelve other community foundations registered double-digit gains in giving in the latest year, while two showed increases of less than 10 percent. On the downside, ten community foundations lowered their levels of giving in 2003. The steepest decline was reported by the Greater Kansas City

TABLE 23. Analysis of Community Foundations by Total Giving Range, 2003

Total Giving Range	Number of Foundations	%	Total Giving[1]	%
$100 million+	1	0.1	$ 117,992	4.7
$25 million–$100 million	28	4.0	1,278,179	50.5
$10 million–$25 million	25	3.6	403,310	15.9
$1 million–$10 million	197	28.2	615,026	24.3
$100,000–$1 million	281	40.2	110,755	4.4
Under $100,000	167	23.9	6,268	0.2
Total	**699**	**100.0**	**$2,531,530**	**100.0**

Source: The Foundation Center, *Foundation Yearbook*, 2005. Dollars in thousands.
[1]Includes grants, scholarships, and employee matching gifts; excludes set-asides, loans, PRIs, and program expenses.

Community Foundation, with giving down 25.4 percent from $98.5 million to $73.4 million.

To qualify for ranking among the top 25 community foundations by total giving, the minimum spending level was $30 million, distributed by the Greater Cincinnati Foundation (OH), compared to the maximum of $118 million, awarded by the New York Community Trust. In addition to the Trust, eleven other funders distributed at least $50 million. Among the top five funders in 2003, the threshold for giving was $73.4 million, compared with $64.5 million in the prior year.

2003 Assets

- **Community foundations' assets grew nearly 15 percent**

- **Assets totaled $34.2 billion, more than double the value in 1996**

Change in Community Foundation Assets. Assets of community foundations rose 14.7 percent in 2003, exceeding the asset gains reported by independent (9.6 percent), corporate (7.1 percent), and operating (4.2 percent) foundations (Table 16). Assets reached a record $34.2 billion in the latest year, up from $29.8 billion in 2002. This also represented a more than doubling of assets from $15.9 billion in 1996.

Assets of Large vs. Small Community Foundations. In 2003, 318 community foundations held assets of $10 million or over, up 8.2 percent from the 294 tracked in the prior year (Table 24). These foundations held 96.5 percent of all community fund assets, practically unchanged from the prior year. A total of 262 funders reported assets ranging from $1 million to $10 million, while 119 had not yet reached the $1 million mark in asset accumulation.

While overall assets of community foundations rose substantially between 2002 and 2003, the biggest increases were concentrated among the largest foundations. For example, total assets of community foundations with endowments of $10 million or more grew 15 percent in 2003. Among foundations in this group,

the biggest asset gains were posted by those with endowments of $250 million to $1 billion (up 28.7 percent). By comparison, total assets of community foundations with assets of under $10 million increased a much more modest 6.9 percent.

Assets of Top 25 Community Foundations. Qualifying asset values among the 25 largest community funds by asset size ranged from the Arizona Community Foundation's $355.3 million to the $1.8 billion held by the New York Community Trust. Aside from the New York Community Trust, three other community foundations held assets of at least $1 billion: the Cleveland Foundation ($1.5 billion), Chicago Community Trust ($1.2 billion), and Marin Community Foundation ($1.1 billion). Collectively, assets of the 25 largest community funds increased 20.8 percent, surpassing the 14.7 percent rise reported for community foundations overall.

Among the 25 largest community funds by asset size, 24 realized asset gains in 2003, compared to only four in the prior year. Twenty-two posted double- or triple-digit increases, led by the Central Indiana Community Foundation, whose assets more than doubled from $161.2 million to $419.3 million. Contributing to this growth were gifts totaling $21.1 million and strong investment performance. Among other foundations with exceptional asset growth was the Tulsa Community Foundation (AZ), whose assets ballooned from $391.8 million to $807.3 million in 2003. The foundation's assets benefited from an infusion of $276.3 million—the largest total gift received by a community foundation in 2003. Asset increases for the 22 remaining foundations reporting asset gains in 2003 ranged from 6 percent to 44 percent.

Reflecting the improved economic climate, only one foundation among the 50 largest community foundations by assets showed a decline in total assets in 2003, compared to twenty-one in the prior year. The San Diego Foundation's (CA) assets decreased 1.9 percent between 2002 and 2003, from $385.9 million to $378.5 million.

TABLE 24.   Analysis of Community Foundations by Asset Range, 2003

Asset Range	Number of Foundations	%	Assets	%	Total Giving[1]	%
$1 billion+	4	0.6	$ 5,594,346	16.4	$ 303,176	12.0
$250 million–$1 billion	29	4.1	14,037,568	41.1	1,068,468	42.2
$50 million–$250 million	78	11.1	8,397,710	24.6	730,849	28.8
$10 million–$50 million	207	29.6	4,932,440	14.4	332,049	13.1
$1 million–$10 million	262	37.5	1,137,285	3.3	88,759	3.5
Under $1 million	119	17.0	53,632	0.1	8,229	0.3
Total	**699**	**100.0**	**$34,152,980**	**100.0**	**$2,531,530**	**100.0**

Source: The Foundation Center, *Foundation Yearbook*, 2005. Dollars in thousands. Due to rounding, figures may not add up.
[1]Includes grants, scholarships, and employee matching gifts; excludes set-asides, loans, PRIs, and program expenses.

Community Foundations New to the Top 25. One foundation was new to the top 25 list of community foundations ranked by assets in 2003—the Central Indiana Community Foundation. The foundation's assets shot up from $161.2 million in 2002 to $419.3 million in the latest year, raising its ranking from 44th to 20th place.

2003 Gifts Received

- Gifts to community foundations increased 9.5 percent, ending two consecutive years of decline

- 78 community funds received gifts totaling at least $10 million

Gifts and Bequests to Community Foundations. Unlike independent foundations, which rely primarily on investment performance to increase assets, community foundations seek to raise their asset base both through the growth of their existing pooled funds and through the cultivation of new donors. Both of these sources benefited from the recovery in the stock market and a stronger economy. Gifts to endowments and pass-through funds grew 9.5 percent, from $3.2 billion in 2002 to $3.5 billion in 2003, following two consecutive years of decline. Moreover, the current level of new gifts into community foundations ranks second only to the $3.8 billion peak recorded in 2000.

The number of community foundations receiving gifts totaling at least $50 million increased from 11 in 2002 to 13 in the latest year, while those with gifts of $10 million or more totaled 78, up from 68. Eleven community foundations based in California received gifts totaling at least $10 million, followed by Texas (six), North Carolina (five), Ohio (four), and Tennessee (four). None of the remaining states reported more than three foundations with gifts in this range.

Largest Gifts Received. For the second consecutive year, the Tulsa Community Foundation received the largest amount of gifts into a community foundation—$276.3 million. The foundation primarily funds nonprofit organizations, including federated giving programs, in northeastern Oklahoma. Three other community foundations received gifts of over $100 million in 2003: the California Community Foundation, with gifts totaling $144.5 million; the Oregon Community Foundation, with gifts totaling $139 million; and the Greater Kansas City Community Foundation, with gifts totaling $111.7 million.

Grantmaking Operating Foundations

Operating foundations primarily conduct their own programs or provide a direct service. Most grantmaking operating foundations make only a tiny number of grants related to their primary mission. However, a few have established distinct grants programs. For various reasons, grantmaking operating foundations experienced very strong growth in the mid-1990s, prompting the Foundation Center to include them in this analysis of trends by foundation type. Since operating foundations are not required to make grants, however, recent trends are not necessarily predictors of future giving.

The dynamic growth of operating foundations in recent years suggests that more donors are choosing to develop and administer their own programs instead of serving as funding resources for other groups. It also suggests that the lines between operating and independent foundations may be blurring, since operating foundations in some cases are becoming large grantmakers.

2003 Giving

- Giving by grantmaking operating foundations reached a record $1.7 billion

- Operating foundation grants represented nearly 6 percent of all foundation giving

Change in Number of Operating Foundations. In 2003, 4,159 operating foundations reported grant activity, up 4.3 percent from last year and four times the 1,040 grantmakers reported for 1987. In the 1980s and early 1990s, the number of grantmaking operating foundations grew faster than all other types of foundations. From 1993 to 1999, operating foundations maintained a roughly consistent 5.5 percent share of the overall number of grantmaking foundations. Since 2000, their share has inched up to approximately 6 percent.

Operating Foundations as a Share of All Giving. Since 1989, operating foundations' share of all foundation grant dollars has been creeping upward, albeit at an uneven pace. In 2003, their share of giving reached 5.8 percent, up from 3.9 percent in the prior year. This record share amounted to more than three times the 1.8 percent share of total giving they represented in 1987, when the Foundation Center began tracking data separately for operating foundations.

Giving by operating foundations jumped 46 percent in 2003—far surpassing the growth in giving by other types of foundations. By dollar amount, their giving reached a record $1.7 billion—over fourteen times greater than the $119 million reported in 1987. Much of this growth was attributable to a small number of foundations established over the past decade, such as the Bristol-Myers Squibb Patient Assistance Foundation (NJ), Janssen Ortho Patient Assistance Foundation (NJ), and the American Contemporary Art Foundation (NY), along with very recently established operating foundations—such as the Merck Patient Assistance Program (NJ) and Boehringer Ingelheim Cares Foundation (CT), both of which were established in 2001.

Top Ten Operating Foundations with Increased Funding. The ten largest operating foundations by giving accounted for nearly 90 percent ($1.5 billion) of the $1.7 billion in grants paid by all grantmaking operating foundations. The three largest by giving—the Bristol-Myers Squibb Patient Assistance Foundation, Merck Patient Assistance Program, and Janssen Ortho Patient Assistance Foundation—were responsible for $996.8 million, or close to three-fifths (57.2 percent) of dollars distributed.

For the third consecutive year, the Bristol-Myers Squibb Patient Assistance Foundation ranked as the top giver among grantmaking operating foundations. The foundation increased its giving 35.2 percent in 2003, from $297.1 million to $401.8 million. Established in 1999, it provides support for low-income patients without prescription drug insurance coverage.

Operating Foundations New to the Top Ten. Four funders were new to the list of top ten grantmaking operating foundations by total giving in 2003, including the second-ranked Merck Patient Assistance Program, fourth-ranked American Contemporary Art Foundation, eighth-ranked Genzyme Charitable Foundation (MA), and tenth-ranked McGregor Foundation (OH).

2003 Assets

• **Operating foundation assets increased over 4 percent**

Change in Operating Foundation Assets. Asset values of the 4,159 grantmaking operating foundations rose 4.2 percent in 2003—far less than the growth in assets reported for community, independent, and corporate foundations (Table 16). This growth followed two consecutive years of decline in the assets of grantmaking operating foundations. Still, the $28 billion in assets held by operating foundations in 2003 represented the second highest level of assets on record, after the $31 billion reported in 2000.

Assets of Large vs. Small Operating Foundations. Of the 4,159 active grantmaking operating foundations, close to one-fourth (23.8 percent) held assets of at least $1 million (Table 25). Yet together these funders controlled 97.7 percent of operating foundations' assets. The top ten operating foundations by assets alone accounted for nearly half (49.3 percent) of the total assets of all grantmaking operating foundations. By far the largest operating foundation by asset size was the J. Paul Getty Trust, with $9.1 billion in assets. Established in 1953 and fully endowed in 1982, Getty's funding focuses on arts conservation and scholarships in the history of art. In 2003, it gave out $19.3 million in grants, down from $21 million in 2002.

2003 Gifts Received

• **Grantmaking operating foundations received $2.3 billion in gifts**

Gifts and Bequests to Operating Foundations. Grantmaking operating foundations received $2.3 billion in gifts and bequests in 2003, up 13.1 percent from the prior year. Five foundations reported over $100 million in gifts from their donors in the latest year, led by the Bristol-Myers Squibb Patient Assistance Foundation, which received $413.6 million, and the Merck Patient Assistance Program, which received $381.5 million.

TABLE 25. Analysis of Grantmaking Operating Foundations by Asset Range, 2003

Asset Range	Number of Foundations	%	Assets	%	Total Giving[1]	%	Program Expenditures	%
$250 million+	15	0.4	$17,489,497	62.5	$ 287,585	16.5	$ 357,919	27.0
$50 million–$250 million	49	1.2	4,586,157	16.4	274,360	15.7	331,119	25.0
$10 million–$50 million	136	3.3	2,793,034	10.0	87,435	5.0	131,743	9.9
$1 million–$10 million	790	19.0	2,464,328	8.8	704,293	40.4	434,073	32.8
Under $1 million	3,169	76.2	641,821	2.3	390,276	22.4	70,322	5.3
Total	**4,159**	**100.0**	**$27,974,838**	**100.0**	**$1,743,950**	**100.0**	**$1,325,178**	**100.0**

Source: The Foundation Center, *Foundation Yearbook,* 2005. Dollars in thousands. Due to rounding, figures may not add up.

[1]For some operating foundations, total giving amount includes grants and program expenses; for others, total giving amount includes only grants. Most operating foundations' qualifying distributions are paid out for administration of operating programs and not for grants.

Endnotes

1. For a more detailed summary of the differences between various types of foundations and the federal regulations that govern them, see the Foundation Center's *Foundation Fundamentals*; see also *The Handbook of Private Foundation*, Washington, DC: Council on Foundations, 2005. See also Appendix B of this report ("Regulation of Private Foundations").
2. Additional data on foundations by type are available in the FC Stats section of the Foundation Center's Web site (see www.fdncenter.org/fc_stats).
3. The F.W. Olin Foundation plans to terminate at the end of 2005. It intends to distribute most of its remaining assets to the Franklin W. Olin College of Engineering.
4. The Foundation Center's count of community foundations falls short of the number identified by other national sources, explained in part by different methods of tracking supporting organizations. The Center reports as one fiscal unit a community foundation and the supporting funds that it administers, if those funds are jointly reported in the foundation's IRS tax form 990 or on the Center's annual survey. The Center's data on active grantmakers also excludes startup community foundations that have not yet awarded grants.

APPENDIX A

Methodology

Foundation Yearbook analyzes and interprets in one volume the entire body of data collected by the Foundation Center on active grantmaking foundations. Its purpose is fourfold: (1) to measure the dimensions of U.S. grantmaking foundations by number, assets, giving, and gifts received as a means of documenting their capacity for supporting the nonprofit sector; (2) to examine differences in the growth and distribution of foundations and their resources by foundation location and type; (3) for the largest foundations, to track growth over time by foundation type; and (4) to monitor new foundation formation.

Foundation Yearbook draws from a range of current and historical data sources. It adheres to underlying foundation definitions and criteria used throughout the Foundation Center's publications. The following methodology discussion examines the universe of philanthropic foundations covered by *Foundation Yearbook*, the sources of data for smaller and larger foundations, and the timeliness of fiscal information.

Counting Foundations: How We Differ from the IRS

The Internal Revenue code defines the category of "private foundation" only by the exclusion of other nonprofit organizations. Under IRS regulations, organizations such as libraries, museums, and homes for the aged may be counted as foundations if they "happen to be endowed by an individual or single family, or if they were established as public charities and lose that status by failing to prove they have received ongoing financial support from the general public."[1]

The Foundation Center supports a narrower private foundation definition: a nongovernmental, nonprofit

organization with its own funds (usually from a single source, either an individual, a family, or a corporation) and program managed by its own trustees and directors, established to maintain or aid educational, social, charitable, religious, or other activities serving the common welfare, primarily by making grants to other nonprofit organizations. Included among these foundations are grantmaking operating foundations (which conduct their own research or direct service programs).

In addition, the Foundation Center tracks community foundations. Like private foundations, these funders make grants, but they receive their funding from the public, generally through contributions received from many donors. (For definitions of foundations by type, see Figure 27 in Chapter 4.)

In counting "active grantmaking foundations," the Foundation Center excludes organizations deemed terminated or inactive and those not intended to fulfill the philanthropic functions the Center ascribes to private foundations. In general, those falling into the latter category are failed public charities.

In early 2005, staff research documented 66,398 active grantmaking foundations. All of these foundations had made grants of at least one dollar in their most current fiscal year on record—which ranged from 2001 to 2004—and had not terminated operations.

Among those foundations filing 990-PF tax returns in the most recent year, but not included in this total, were foundations that did not award any grants (including many newly established funders); and foundations that had either terminated operations, merged into another foundation or corporate giving program, changed status to a public charity, or become inactive.

1. D. Freeman, J. Edie, J. Nober, and the Council on Foundations, *The Handbook of Private Foundations*, Washington, DC: Council on Foundations, 2005.

Sources of Historical Data

To establish time-line data going back to 1975, *Foundation Yearbook* consolidates summary information published by the Foundation Center in the statistical introductions to its major reference publications. Fiscal data summaries were drawn from the *National Data Book of Foundations* (editions 2-16), and the Foundation Center database.

Sources of Current Data

The Foundation Center's sources of data on active grantmaking foundations include:

- the yearly transaction tape produced by the Internal Revenue Service from the annual information returns (Form 990-PF) filed by private foundations during the given period

- CD-ROMs of Form 990-PF that the Center receives monthly from the IRS

- financial and program information provided by foundations in annual reports, other foundation publications, Web sites, or in questionnaires mailed out annually by the Foundation Center to more than 25,000 larger foundations (including those in *The Foundation Directory* and *The Foundation Directory, Part 2*). These questionnaires are the primary source of detailed information on foundation establishment, staffing, and reporting, and of detailed financial reporting on loans and other program-related investments, and grants and scholarships to individuals

Gathering Data on Active Community Foundations

Most community foundations are not included on the IRS tape or in the shipments of 990-PF tax returns since they are classified as public charities and file different information returns (IRS Form 990). Thus, the Center, after working with the Council on Foundations to identify existing community foundations, has gathered information on these foundations primarily through annual surveys. In 2004, the Center received responses from 699 community foundations, and this information was added to the foundation database file by our staff. Most of the community foundations appearing in either of the *Directory* publications also publish an annual report or program statement.

The 2003 Report Year: How Consistent Is the Information?

Because the Foundation Center depends on a range of fiscal reporting sources—some far more timely than others—a single snapshot of the 66,398 active foundations typically contains fiscal data spanning three to four years. By updating foundation listings directly from questionnaires, annual reports, and Form 990-PF, the Center has been able to improve the timeliness of data by at least 50 percent over that reported on the IRS transaction tape. For this analysis, 75 percent of the 66,398 foundation listings contained 2003 or early 2004 fiscal data, 18 percent contained 2002 data, and the balance contained 2001 data.

Those 49,707 foundations reporting 2003 or early 2004 fiscal information accounted for 86 percent of the total assets and 80 percent of the total giving reported in this edition. An additional nearly 12 percent of assets and 18 percent of grant dollars reflected 2002 fiscal data. Finally, 2 percent of asset and grants information reflected 2001 fiscal data.

Analyses of Larger Foundations

Throughout this report, expanded analyses have been conducted on foundations holding assets of at least $1 million or making grants totaling $100,000 or more. A total of close to 21,000 private and community foundations met this standard for report year 2003.

The special reports on these foundations include foundation formation by decade. Historical data were taken from *The Foundation Directory*, while more recent fiscal data came directly from the Center's database.

Tracking New Foundations

Reporting on new foundation formation is an important and unique feature of *Foundation Yearbook*. The Center has researched and maintained comprehensive records on foundation establishment for the larger foundations since the 1970s, allowing us to chart the growth of foundations across decades. Since 1987, when the Center began to process data directly from the 990-PF tax return, our staff has researched formation year data for small foundations. As a result, in early 2005, we were able to identify 43,294 active foundations established since 1980.

These new foundations represent nearly 98 percent of the 44,310 foundation records added to the Center's database since 1980. Others are not new foundations but rather "private foundation" non-exempt charitable trusts (NECTs). Starting around 1982, the IRS required that these trusts file Form 990-PF.

Foundation Giving Forecast Survey

Giving estimates for 2004 are based on figures reported by over 900 large and mid-size independent, corporate, and community foundations in the Foundation Center's 2005 "Foundation Giving Forecast Survey" combined with year-end fiscal indicators. Survey respondents accounted for 31 percent of total estimated giving by independent, corporate, and community foundations in 2004. By foundation type, the share of estimated 2004 giving was higher for the 142 community foundation respondents in the set (51 percent), roughly equal for the 704 independent foundation respondents (30 percent), and lower for the 99 corporate foundation respondents (23 percent). The assessment of prospects for 2005 giving was based on information reported in the survey by 871 foundations, including 644 independent, 92 corporate, and 135 community foundations.

APPENDIX B

Regulation of Private Foundations

Nonprofit organizations that are defined as private foundations by the IRS must be organized and operated under specific regulations, which are primarily derived from the Tax Reform Act of 1969, with some significant modifications as a result of the 1976 Tax Reform Act, the Economic Recovery Tax Act of 1981 (ERTA), and tax legislation enacted in 1984, 1986, 1987, and 1988. The Tax Reform Act of 1969 established the criteria for distinguishing private foundations from public charities and set up a separate category of private foundations with a more favorable tax status for operating foundations.

Provisions of the 1969 Tax Reform Act. The Act set forth special rules that, among other things, prohibit certain transactions between foundations and "disqualified persons," including their donors and managers, as "self-dealing"; restrict foundation ownership and control of private business; limit the percentage of an individual's annual income that can be donated to a private foundation as a tax-deductible contribution; and regulate foundation giving to individuals, other foundations, non-exempt organizations, and activities that influence legislation or political campaigns.

To offset costs incurred by the IRS in regulating tax-exempt organizations, the 1969 Act imposed an excise tax on foundations at the rate of 4 percent of net investment income. Because this excise tax generated revenues that greatly exceeded the annual budget of the exempt organizations branch of the IRS, the Revenue Act of 1978 reduced the tax to 2 percent. With tax revenues still many times greater than regulatory costs, the Deficit Reduction Act of 1984 reduced the tax from 2 to 1 percent for any foundation that, in a given tax year, makes sufficient qualifying distributions to meet certain requirements.

Payout Requirements. The 1969 Act also required a private foundation to distribute for charitable purposes, either all of its "adjusted net income" or a percentage, to be set each year, of the market value of that year's assets, whichever was higher. This "payout requirement" was later set at 5 percent of the average market value of that year's assets, to be paid out in the following year. Tax legislation in 1984 limited to .65 percent of net investment assets the amount of administrative expenses incurred in making grants that could be counted as part of the 5 percent payout requirement. The intention was to assure that private foundations could not meet the payout requirement by paying excessive trustee fees or similar non-charitable distributions. The ".65 percent rule" was allowed to lapse at the end of 1990, following a study by the Treasury Department that concluded that it was complex and burdensome and that there was no evidence of excessive administrative expenses on the part of the vast majority of foundations.

Reporting Requirements. The 1969 legislation called for foundations to file two annual information returns with the IRS (Form 990-AR and 990-PF), to make the forms available for inspection by the public, and to file two copies of the forms with state authorities in the state where the foundation is incorporated and maintains its principal offices. Beginning in 1982, foundations have been required to file only one form, a revised version of the Form 990-PF that incorporates all information previously required on the separate forms, and calls for additional information on the foundation's grantmaking policies.

Legislation passed in 1998 and regulations finalized in early 2000 now require private foundations to provide "take home" copies of their current Form 990-PF to any individual who makes a request in person or in writing. These regulations apply to all returns filed after March 13, 2000. Foundations are required to make each annual return filed after this date available for three years. If a foundation's Form

990-PF is widely available through the Internet in a manner that complies with the regulations, it will have satisfied its disclosure requirements and need not provide take home copies.

Other Provisions. The Tax Reform Act of 1969 further prohibits or circumscribes a number of activities on the part of private foundations, including attempts to influence legislation, intervention in political campaigns, and making grants to individuals or to other private foundations or non-501(c)(3) organizations. Making grants in these areas entails a certain amount of planning and paperwork for foundations. For instance, a foundation must have an IRS-approved individual grants program in order to make grants to individuals for study or travel purposes, and must follow up such grants with special reports. A foundation may make a grant for a charitable purpose to an organization that does not qualify as a public charity so long as it exercises "expenditure responsibility," which entails a pre-grant inquiry, a grant contract and follow-up reports to the IRS as part of the 990-PF. Foundations may even support groups that include lobbying or voter education and registration in their activities, provided that they and their grantee follow the regulations governing these areas. More recent directives from the Treasury outline procedures that foundations may follow in connection with grant-making to foreign organizations.

State Regulation. The IRS is responsible for enforcing the federal regulations on foundations, and it accomplishes this primarily through review of the 990-PF reports and through audits of individual foundations. At the state level, regulation is typically the concern of the state attorney general, who enforces the state's not-for-profit corporation law and any statutory or common laws governing charitable trustees. The federal regulatory system was designed to encourage more active state regulation of foundations. The 990-PFs must, for instance, be filed with the appropriate state authorities as well as with the IRS. Thus, in many states, the office of the attorney general serves as another resource for those interested in researching a foundation located in that state.

APPENDIX C

Foundation Center Cooperating Collections
Free Funding Information Centers

The Foundation Center's mission is to strengthen the nonprofit sector by advancing knowledge about U.S. philanthropy. An authoritative source of information on foundation and corporate giving, we ensure free public access to a wide variety of services and comprehensive resources on foundations and grants through our five library/learning centers and a national network of Cooperating Collections. Cooperating Collections are libraries, community foundations, and other nonprofit agencies that make accessible a collection of Foundation Center print and electronic resources, as well as a variety of supplementary materials and educational programs in areas useful to grantseekers. The collection includes:

FOUNDATION DIRECTORY ONLINE or FC SEARCH: THE
   FOUNDATION CENTER'S DATABASE ON CD-ROM
THE FOUNDATION DIRECTORY, PART 2, and SUPPLEMENT
FOUNDATION FUNDAMENTALS
THE FOUNDATION 1000
FOUNDATIONS TODAY SERIES

FOUNDATION GRANTS TO INDIVIDUALS
THE FOUNDATION CENTER'S GUIDE TO GRANTSEEKING
   ON THE WEB
THE FOUNDATION CENTER'S GUIDE TO PROPOSAL
   WRITING

THE FOUNDATION CENTER'S GUIDE TO WINNING
   PROPOSALS
GUIDE TO U.S. FOUNDATIONS, THEIR TRUSTEES,
   OFFICERS, AND DONORS
NATIONAL DIRECTORY OF CORPORATE GIVING
NATIONAL GUIDE TO FUNDING IN. . . . (SERIES)

All five Foundation Center libraries provide free access to both *The Foundation Directory Online* and *FC: Search: The Foundation Center's Database on CD-ROM*. All Cooperating Collections provide access to either the online service or the CD-ROM, and all provide Internet access. Those seeking information on fundraising and nonprofit management can also refer to our Web site (www.fdncenter.org) for a wealth of data and advice on grantseeking, including links to foundation IRS information returns (990-PFs). Because the Cooperating Collections vary in their hours, it is recommended that you call the collection in advance of a visit. To check on new locations or current holdings, call toll-free 1-800-424-9836 or visit www.fdncenter.org/collections/.

FOUNDATION CENTER LIBRARY/LEARNING CENTERS

THE FOUNDATION CENTER
2nd Floor
79 Fifth Ave.
New York, NY 10003
(212) 620-4230

THE FOUNDATION CENTER
312 Sutter St., Suite 606
San Francisco, CA 94108
(415) 397-0902

THE FOUNDATION CENTER
1627 K St., NW, 3rd floor
Washington, DC 20006
(202) 331-1400

THE FOUNDATION CENTER
Kent H. Smith Library
1422 Euclid Ave., Suite 1600
Cleveland, OH 44115
(216) 861-1933

THE FOUNDATION CENTER
Suite 150, Grand Lobby
Hurt Bldg., 50 Hurt Plaza
Atlanta, GA 30303
(404) 880-0094

COOPERATING COLLECTIONS

ALABAMA

BIRMINGHAM PUBLIC LIBRARY
Government Documents
2100 Park Place
Birmingham 35203
(205) 226-3620

HUNTSVILLE PUBLIC LIBRARY
915 Monroe St.
Huntsville 35801
(256) 532-5940

MOBILE PUBLIC LIBRARY
West Regional Library
5555 Grelot Rd.
Mobile 36609-3643
(251) 340-8555

AUBURN UNIVERSITY AT
MONTGOMERY LIBRARY
74-40 East Dr.
Montgomery 36117-3596
(334) 244-3200

ALASKA

CONSORTIUM LIBRARY
3211 Providence Dr.
Anchorage 99508
(907) 786-1848

JUNEAU PUBLIC LIBRARY
292 Marine Way
Juneau 99801
(907) 586-5267

ARIZONA

FLAGSTAFF CITY-COCONINO
COUNTY PUBLIC LIBRARY
300 W. Aspen Ave.
Flagstaff 86001
(928) 779-7670

PHOENIX PUBLIC LIBRARY
Information Services Dept.
1221 N. Central Ave.
Phoenix 85004
(602) 262-4636

TUCSON PIMA PUBLIC LIBRARY
101 N. Stone Ave.
Tucson 87501
(520) 791-4393

ARKANSAS

UNIVERSITY OF ARKANSAS—FT.
SMITH
BOREHAM LIBRARY
5210 Grand Ave.
Ft. Smith 72913
(479) 788-7204

CENTRAL ARKANSAS
LIBRARY SYSTEM
100 Rock St.
Little Rock 72201
(501) 918-3000

CALIFORNIA

KERN COUNTY LIBRARY
Beale Memorial Library
701 Truxtun Ave.
Bakersfield 93301
(661) 868-0755

HUMBOLDT AREA FOUNDATION
Rooney Resource Center
373 Indianola
Bayside 95524
(707) 442-2993

VENTURA COUNTY COMMUNITY
FOUNDATION
Resource Center for Nonprofit
   Organizations

1317 Del Norte Rd., Suite 150
Camarillo 93010
(805) 988-0196

FRESNO REGIONAL FOUNDATION
Nonprofit Advancement Center
3425 N. First St., Suite 101
Fresno 93726
(559) 226-0216

CENTER FOR NONPROFIT
MANAGEMENT IN SOUTHERN
CALIFORNIA
Nonprofit Resource Library
606 South Olive St. #2450
Los Angeles 90014
(213) 623-7080

LOS ANGELES PUBLIC LIBRARY
Mid-Valley Regional Branch Library
16244 Nordhoff St.
North Hills 91343
(818) 895-3654

PHILANTHROPY RESOURCE CENTER
Flintridge Foundation
1040 Lincoln Ave, Suite 100
Pasadena 91103
(626) 449-0839

CENTER FOR NONPROFIT
RESOURCES
Shasta Regional Community
Foundation's Center
Bldg. C, Suite A
2280 Benton Dr.
Redding   96003
(530) 244-1219

RICHMOND PUBLIC LIBRARY
352 Civic Center Plaza
Richmond   94804
(510) 620-6561

RIVERSIDE CITY PUBLIC LIBRARY
3581 Mission Inn Ave.
Riverside   92501
(909) 826-5201

NONPROFIT RESOURCE CENTER
Sacramento Public Library
828 I St., 2nd Floor
Sacramento   95814
(916) 264-2772

SAN DIEGO FOUNDATION
Funding Information Center
1420 Kettner Blvd., Suite 500
San Diego   92101
(619) 235-2300

COMPASSPOINT NONPROFIT
SERVICES
Nonprofit Development Library
1922 The Alameda, Suite 212
San Jose   95126
(408) 248-9505

LOS ANGELES PUBLIC LIBRARY
San Pedro Regional Branch
931 S. Gaffey St.
San Pedro   90731
(310) 548-7779

VOLUNTEER CENTER OF GREATER
ORANGE COUNTY
Nonprofit Resource Center
1901 E. 4th St., Suite 100
Santa Ana   92705
(714) 953-5757

SANTA BARBARA PUBLIC LIBRARY
40 E. Anapamu St.
Santa Barbara   93101-1019
(805) 962-7653

SANTA MONICA PUBLIC LIBRARY
1324 Fifth St.
Santa Monica   90401
(310) 458-8600

SONOMA COUNTY LIBRARY
3rd & E Sts.
Santa Rosa   95404
(707) 545-0831

SEASIDE BRANCH LIBRARY
550 Harcourt Ave.
Seaside   93955
(831) 899-8131

SIERRA NONPROFIT SUPPORT
CENTER
39 No. Washington St. #F
Sonora   95370-0905
(209) 533-1093

COLORADO

PENROSE LIBRARY
20 N. Cascade Ave.
Colorado Springs 80903
(719) 531-6333

DENVER PUBLIC LIBRARY
10 W. 14th Ave. Pkwy.
Denver   80204
(720) 865-1111

PUEBLO CITY-COUNTY LIBRARY
DISTRICT
100 East Abriendo Ave.
Pueblo   81004-4232
(719) 562-5600

CONNECTICUT

DANBURY PUBLIC LIBRARY
170 Main St.
Danbury   06810
(203) 797-4527

GREENWICH LIBRARY
101 W. Putnam Ave.
Greenwich   06830
(203) 622-7900

HARTFORD PUBLIC LIBRARY
500 Main St.
Hartford   06103
(860) 695-6300

NEW HAVEN FREE PUBLIC LIBRARY
133 Elm St.
New Haven 06510-2057
(203) 946-7431

DELAWARE

UNIVERSITY OF DELAWARE
Hugh Morris Library
181 South College Ave.
Newark   19717-5267
(302) 831-2432

FLORIDA

BARTOW PUBLIC LIBRARY
2151 S. Broadway Ave.
Bartow 33830
(863) 534-0931

VOLUSIA COUNTY LIBRARY CENTER
City Island
105 E. Magnolia Ave.
Daytona Beach   32114-4484
(386) 257-6036

NOVA SOUTHEASTERN UNIVERSITY
Library, Research and Information
Technology Center
3100 Ray Ferrero Jr. Blvd.
Fort Lauderdale   33314
(954) 262-4613

INDIAN RIVER COMMUNITY
COLLEGE
Learning Resources Center
3209 Virginia Ave.
Fort Pierce   34981-5596
(561) 462-4757

JACKSONVILLE PUBLIC LIBRARIES
Grants Resource Center
122 N. Ocean St.
Jacksonville   32202
(904) 630-2665

MIAMI-DADE PUBLIC LIBRARY
Humanities/Social Science
101 W. Flagler St.
Miami   33130
(305) 375-5575

ORANGE COUNTY LIBRARY SYSTEM
Social Sciences Dept.
101 E. Central Blvd.
Orlando   32801
(407) 835-7323

SELBY PUBLIC LIBRARY
Reference
1331 1st St.
Sarasota   34236
(941) 861-1100

STATE LIBRARY OF FLORIDA
R.A. Gray Bldg.
Tallahassee 32399-0250
(850) 245-6600

HILLSBOROUGH COUNTY PUBLIC
LIBRARY COOPERATIVE
John F. Germany Public Library
900 N. Ashley Dr.
Tampa   33602
(813) 273-3652

COMMUNITY FOUNDATION OF
PALM BEACH & MARTIN COUNTIES
700 S. Dixie Hwy., Suite 200
West Palm Beach   33401
(561) 659-6800

GEORGIA

HALL COUNTY LIBRARY SYSTEM
127 Main St. NW
Gainesville   30501
(770) 532-3311

WASHINGTON MEMORIAL LIBRARY
1180 Washington Ave.
Macon   31201
(478) 744-0828

THOMAS COUNTY PUBLIC LIBRARY
201 N. Madison St.
Thomasville   31792
(229) 225-5252

HAWAII

UNIVERSITY OF HAWAII
Hamilton Library
2550 The Mall
Honolulu   96822
(808) 956-7214

IDAHO

BOISE PUBLIC LIBRARY
Funding Information Center
715 S. Capitol Blvd.
Boise   83702
(208) 384-4024

CALDWELL PUBLIC LIBRARY
1010 Dearborn St.
Caldwell   83605
(208) 459-3242

MARSHALL PUBLIC LIBRARY
113 South Garfield
Pocatello   83204
(208) 232-1263

ILLINOIS

CARBONDALE PUBLIC LIBRARY
405 West Main St.
Carbondale   62901
(618) 457-0354

DONORS FORUM OF CHICAGO
208 S. LaSalle, Suite 735
Chicago   60604
(312) 578-0175

EVANSTON PUBLIC LIBRARY
1703 Orrington Ave.
Evanston   60201
(847) 866-0300

ROCK ISLAND PUBLIC LIBRARY
401 19th St.
Rock Island   61201-8143
(309) 732-7323

UNIVERSITY OF ILLINOIS
AT SPRINGFIELD, LIB 140
Brookens Library
One University Plaza
Springfield   62703-5407
(217) 206-6633

INDIANA

EVANSVILLE–VANDERBURGH
PUBLIC LIBRARY
22 SE 5th St.
Evansville   47708
(812) 428-8200

ALLEN COUNTY PUBLIC LIBRARY
200 East Berry St.
Ft. Wayne   46802
(260) 421-1238

INDIANAPOLIS–MARION COUNTY
PUBLIC LIBRARY
202 North Alabama
Indianapolis   46206
(317) 269-1700

VIGO COUNTY PUBLIC LIBRARY
1 Library Square
Terre Haute   47807
(812) 232-1113

VALPARAISO UNIVERSITY
Moellering Library
1509 Chapel Dr.
Valparaiso   46383
(219) 464-5364

IOWA

CEDAR RAPIDS PUBLIC LIBRARY
500 1st St., SE
Cedar Rapids   52401
(319) 398-5123

SOUTHWESTERN COMMUNITY
COLLEGE
Learning Resource Center
1501 W. Townline Rd.
Creston   50801
(641) 782-7081

DES MOINES PUBLIC LIBRARY
100 Locust St.
Des Moines   50309-1791
(515) 283-4152

SIOUX CITY PUBLIC LIBRARY
Siouxland Funding Research Center
529 Pierce St.
Sioux City   51101-1203
(712) 255-2933

KANSAS

PIONEER MEMORIAL LIBRARY
375 West 4th St.
Colby   67701
(785) 462-4470

DODGE CITY PUBLIC LIBRARY
1001 2nd Ave.
Dodge City   67801
(620) 225-0248

KEARNY COUNTY LIBRARY
101 East Prairie
Lakin   67860
(620) 355-6674

SALINA PUBLIC LIBRARY
301 West Elm
Salina 67401
(785) 825-4624

TOPEKA AND SHAWNEE COUNTY
PUBLIC LIBRARY
1515 SW 10th Ave.
Topeka   66604
(785) 580-4400

WICHITA PUBLIC LIBRARY
223 S. Main St.
Wichita   67202
(316) 261-8500

KENTUCKY

WESTERN KENTUCKY UNIVERSITY
Helm-Cravens Library
110 Helm Library
Bowling Green   42101-3576
(270) 745-6163

LEXINGTON PUBLIC LIBRARY
140 E. Main St.
Lexington   40507-1376
(859) 231-5520

LOUISVILLE FREE PUBLIC LIBRARY
301 York St.
Louisville   40203
(502) 574-1617

LOUISIANA

EAST BATON ROUGE PARISH
LIBRARY
Centroplex Branch Grants Collection
120 St. Louis St.
Baton Rouge   70802
(225) 389-4967

BEAUREGARD PARISH LIBRARY
205 S. Washington Ave.
De Ridder 70634
(337) 463-6217

OUACHITA PARISH PUBLIC LIBRARY
1800 Stubbs Ave.
Monroe 71201
(318) 327-1490

NEW ORLEANS PUBLIC LIBRARY
Business & Science Division
219 Loyola Ave.
New Orleans 70112
(504) 596-2580

SHREVE MEMORIAL LIBRARY
424 Texas St.
Shreveport 71120-1523
(318) 226-5894

MAINE

UNIVERSITY OF SOUTHERN
MAINE LIBRARY
Maine Philanthropy Center
314 Forrest Ave.
Portland 04104-9301
(207) 780-5029

MARYLAND

ENOCH PRATT FREE LIBRARY
Social Science & History Dept.
400 Cathedral St.
Baltimore 21201
(410) 396-5320

MASSACHUSETTS

ASSOCIATED GRANT MAKERS OF
MASSACHUSETTS
55 Court St.
Suite 520
Boston 02108
(617) 426-2606

BOSTON PUBLIC LIBRARY
Soc. Sci. Reference
700 Boylston St.
Boston 02116
(617) 536-5400

WESTERN MASSACHUSETTS
FUNDING RESOURCE CENTER
65 Elliot St.
Springfield 01101-1730
(413) 452-0697

WORCESTER PUBLIC LIBRARY
Grants Resource Center
3 Salem Square
Worcester 01608
(508) 799-1655

MICHIGAN

ALPENA COUNTY LIBRARY
211 N. 1st St.
Alpena 49707
(989) 356-6188

UNIVERSITY OF MICHIGAN
Graduate Library
Reference & Research Services Dept.
209 Hatcher N.
Ann Arbor 48109-1205
(734) 763-1539

WILLARD PUBLIC LIBRARY
Nonprofit & Funding Resource
Collections
7 W. Van Buren St.
Battle Creek 49017
(269) 969-2100

HENRY FORD CENTENNIAL LIBRARY
16301 Michigan Ave.
Dearborn 48126
(313) 943-2330

WAYNE STATE UNIVERSITY
134 Purdy/Kresge Library
Detroit 48202
(513) 577-6424

MICHIGAN STATE UNIVERSITY
LIBRARIES
Main Library
Funding Center
100 Library
East Lansing 48824-1048
(517) 432-6123

FARMINGTON COMMUNITY
LIBRARY
32737 W. 12 Mile Rd.
Farmington Hills 48334
(248) 553-0300

UNIVERSITY OF MICHIGAN—FLINT
Frances Willson Thompson Library
Flint 48502-1950
(810) 762-3413

GRAND RAPIDS PUBLIC LIBRARY
111 Library St. NE
Grand Rapids 49503-3268
(616) 988-5400

MICHIGAN TECHNOLOGICAL
UNIVERSITY
Harold Meese Center, Corporate
Services
1400 Townsend Dr.
Houghton 49931-1295
(906) 487-2228

WEST SHORE COMMUNITY
COLLEGE LIBRARY
3000 North Stiles Rd.
Scottville 49454-0277
(231) 845-6211

TRAVERSE AREA DISTRICT LIBRARY
610 Woodmere Ave.
Traverse City 49686
(231) 932-8500

MINNESOTA

BRAINERD PUBLIC LIBRARY
416 South Fifth St.
Brainerd 56401
(218) 829-5574

DULUTH PUBLIC LIBRARY
520 W. Superior St.
Duluth 55802
(218) 723-3802

SOUTHWEST STATE UNIVERSITY
University Library
N. Hwy. 23
Marshall 56253
(507) 537-6108

MINNEAPOLIS PUBLIC LIBRARY
250 Marquette Ave.
Minneapolis 55401
(612) 630-6000

ROCHESTER PUBLIC LIBRARY
101 2nd St. SE
Rochester 55904-3777
(507) 285-8002

ST. PAUL PUBLIC LIBRARY
90 W. Fourth St.
St. Paul 55102
(651) 266-7000

MISSISSIPPI

LIBRARY OF HATTIESBURG, PETAL
AND FORREST COUNTY
329 Hardy St.
Hattiesburg 39401-3824
(601) 582-4461

JACKSON/HINDS LIBRARY SYSTEM
300 N. State St.
Jackson 39201
(601) 968-5803

MISSOURI

COUNCIL ON PHILANTHROPY
University of Missouri—Kansas City
Center for Business Innovation
4747 Troost, #207
Kansas City 64171-0813
(816) 235-1176

KANSAS CITY PUBLIC LIBRARY
14 West 10th St.
Kansas City 64105-1702
(816) 701-3400

ST. LOUIS PUBLIC LIBRARY
1301 Olive St.
St. Louis 63103
(314) 241-2288

SPRINGFIELD-GREENE
COUNTY LIBRARY
The Library Center
4653 S. Campbell
Springfield 65807
(417) 874-8110

MONTANA

FALLON COUNTY LIBRARY
6 West Fallon Ave.
Baker 59313-1037
(406)778-7160

MONTANA STATE
UNIVERSITY—BILLINGS
Library—Special Collections
1500 N. 30th St.
Billings 59101-0245
(406) 657-1687

BOZEMAN PUBLIC LIBRARY
220 E. Lamme
Bozeman 59715
(406) 582-2402

LINCOLN COUNTY PUBLIC
LIBRARIES
Libby Public Library
220 West 6th St.
Libby 59923
(406) 293-2778

UNIVERSITY OF MONTANA
Mansfield Library
32 Campus Dr. #9936
Missoula 59812-9936
(406) 243-6800

NEBRASKA

UNIVERSITY OF NEBRASKA—
LINCOLN
14th & R Sts.
Lincoln 68588-2848
(402) 472-2848

OMAHA PUBLIC LIBRARY
W. Dale Clark Library
Social Sciences Dept.
215 S. 15th St.
Omaha 68102
(402) 444-4826

NEVADA

GREAT BASIN COLLEGE LIBRARY
1500 College Pkwy.
Elko 89801
(775) 753-2222

CLARK COUNTY LIBRARY
1401 E. Flamingo
Las Vegas 89119
(702) 507-3400

WASHOE COUNTY LIBRARY
301 S. Center St.
Reno 89501
(775) 327-8300

NEW HAMPSHIRE

CONCORD PUBLIC LIBRARY
45 Green St.
Concord 03301
(603) 225-8670

PLYMOUTH STATE UNIVERSITY
Herbert H. Lamson Library
Plymouth 03264
(603) 535-2258

NEW JERSEY

CUMBERLAND COUNTY LIBRARY
800 E. Commerce St.
Bridgeton 08302
(856) 453-2210

FREE PUBLIC LIBRARY OF ELIZABETH
11 S. Broad St.
Elizabeth 07202
(908) 354-6060

NEWARK ENTERPRISE COMMUNITY
RESOURCE DEVELOPMENT CENTER
303-309 Washington St., 5th Floor
Newark 07102
(973) 624-8300

COUNTY COLLEGE OF MORRIS
Learning Resource Center
214 Center Grove Rd.
Randolph 07869
(973) 328-5296

NEW JERSEY STATE LIBRARY
185 W. State St.
Trenton 08625-0520
(609) 292-6220

NEW MEXICO

ALBUQUERQUE/BERNALILLO
COUNTY LIBRARY SYSTEM
501 Copper Ave. NW
Albuquerque 87102
(505) 768-5141

NEW MEXICO STATE LIBRARY
Information Services
1209 Camino Carlos Rey
Santa Fe 87507
(505) 476-9702

NEW YORK

NEW YORK STATE LIBRARY
Humanities Reference
Cultural Education Center, 6th Floor
Empire State Plaza
Albany 12230
(518) 474-5355

BROOKLYN PUBLIC LIBRARY
Society, Science and Technology
Division
Grand Army Plaza
Brooklyn 11238
(718) 230-2122

BUFFALO & ERIE COUNTY
PUBLIC LIBRARY
Business, Science & Technology Dept.
1 Lafayette Square
Buffalo 14203-1887
(716) 858-7097

SOUTHEAST STEUBEN
COUNTY LIBRARY
300 Nasser Civic Center Plaza
Corning 14830
(607) 936-3713

HUNTINGTON PUBLIC LIBRARY
338 Main St.
Huntington 11743
(631) 427-5165

QUEENS BOROUGH PUBLIC
LIBRARY
Social Sciences Division
89-11 Merrick Blvd.
Jamaica 11432
(718) 990-0700

LEVITTOWN PUBLIC LIBRARY
1 Bluegrass Ln.
Levittown   11756
(516) 731-5728

ADRIANCE MEMORIAL LIBRARY
Special Services Dept.
93 Market St.
Poughkeepsie   12601
(914) 485-3445

THE RIVERHEAD FREE LIBRARY
330 Court St.
Riverhead 11901
(631) 727-3228

ROCHESTER PUBLIC LIBRARY
Social Sciences
115 South Ave.
Rochester   14604
(585) 428-8120

ONONDAGA COUNTY PUBLIC
LIBRARY
447 S. Salina St.
Syracuse   13202-2494
(315) 435-1900

UTICA PUBLIC LIBRARY
303 Genesee St.
Utica   13501
(315) 735-2279

WHITE PLAINS PUBLIC LIBRARY
100 Martine Ave.
White Plains   10601
(914) 422-1480

YONKERS PUBLIC LIBRARY
Riverfront Library
One Larkin Center
Yonkers   10701
(914) 337-1500

NORTH CAROLINA

PACK MEMORIAL LIBRARY
Community Foundation of Western
North Carolina
67 Haywood St.
Asheville   28802
(828) 254-4960

THE DUKE ENDOWMENT
100 N. Tryon St., Suite 3500
Charlotte   28202-4012
(704) 376-0291

DURHAM COUNTY PUBLIC LIBRARY
300 N. Roxboro St.
Durham   27702
(919) 560-0100

NEW HANOVER COUNTY PUBLIC
LIBRARY
201 Chestnut St.
Wilmington   28401-3942
(910) 798-6301

FORSYTH COUNTY PUBLIC LIBRARY
660 W. 5th St.
Winston-Salem   27101
(336) 727-2264

NORTH DAKOTA

BISMARCK PUBLIC LIBRARY
515 N. 5th St.
Bismarck   58501-4081
(701) 222-6410

FARGO PUBLIC LIBRARY
102 N. 3rd St.
Fargo   58102
(701) 241-1491

MINOT PUBLIC LIBRARY
516 Second Ave. SW
Minot   58701-3792
(701) 852-1045

OHIO

STARK COUNTY DISTRICT LIBRARY
715 Market Ave. N.
Canton   44702
(330) 452-0665

PUBLIC LIBRARY OF CINCINNATI &
HAMILTON COUNTY
Grants Resource Center
800 Vine St.—Library Square
Cincinnati   45202-2071
(513) 369-6000

COLUMBUS METROPOLITAN
LIBRARY
Business and Technology
96 S. Grant Ave.
Columbus   43215
(614) 645-2590

DAYTON METRO LIBRARY
Grants Information Center
215 E. Third St.
Dayton   45402
(937) 227-9500

MANSFIELD/RICHLAND COUNTY
PUBLIC LIBRARY
43 W. 3rd St.
Mansfield   44902
(419) 521-3110

PORTSMOUTH PUBLIC LIBRARY
1220 Gallia St.
Portsmouth   45662
(740) 354-5688

TOLEDO–LUCAS COUNTY
PUBLIC LIBRARY
325 Michigan St.
Toledo   43612
(419) 259-5207

PUBLIC LIBRARY OF
YOUNGSTOWN & MAHONING
COUNTY
305 Wick Ave.
Youngstown   44503
(330) 744-8636

OKLAHOMA

OKLAHOMA CITY UNIVERSITY
Dulaney Browne Library
2501 N. Blackwelder
Oklahoma City   73106
(405) 521-5822

TULSA CITY–COUNTY LIBRARY
400 Civic Center
Tulsa   74103
(918) 596-7977

OREGON

OREGON INSTITUTE OF
TECHNOLOGY
Library
3201 Campus Dr.
Klamath Falls   97601-8801
(541) 885-1770

MULTNOMAH COUNTY LIBRARY
801 SW 10th Ave.
Portland   97205
(503) 988-5123

OREGON STATE LIBRARY
State Library Bldg.
250 Winter St. NE
Salem   97301-3950
(503) 378-4277

PENNSYLVANIA

NORTHAMPTON COMMUNITY
COLLEGE
The Paul and Harriet Mack Library
3835 Green Pond Rd.
Bethlehem   18017
(610) 861-5360

ERIE COUNTY LIBRARY SYSTEM
160 E. Front St.
Erie   16507
(814) 451-6927

DAUPHIN COUNTY LIBRARY
SYSTEM
East Shore Area Library
4501 Ethel St.
Harrisburg   17109
(717) 652-9380

HAZLETON AREA PUBLIC LIBRARY
55 North Church St.
Hazleton 18201
(570) 454-2961

LANCASTER COUNTY LIBRARY
125 N. Duke St.
Lancaster   17602
(717) 394-2651

FREE LIBRARY OF PHILADELPHIA
Regional Foundation Center
1901 Vine St.
Philadelphia   19103-1189
(215) 686-5423

CARNEGIE LIBRARY OF PITTSBURGH
Foundation Collection
612 Smithfield St.
Pittsburgh   15222-1818
(412) 281-7143

POCONO NORTHEAST
DEVELOPMENT FUND
James Pettinger Memorial Library
1151 Oak St.
Pittston   18640
(570) 655-5581

READING PUBLIC LIBRARY
100 S. 5th St.
Reading   19602
(610) 655-6355

JAMES V. BROWN LIBRARY
19 East Fourth St.
Williamsport   17701
(570) 326-0536

MARTIN LIBRARY
159 E. Market St.
York   17401
(717) 846-5300

RHODE ISLAND

PROVIDENCE PUBLIC
LIBRARY
225 Washington St.
Providence   02906
(401) 455-8088

SOUTH CAROLINA

ANDERSON COUNTY LIBRARY
300 N. McDuffie St.
Anderson   29622
(864) 260-4500

CHARLESTON COUNTY LIBRARY
68 Calhoun St.
Charleston   29401
(843) 805-6930

SOUTH CAROLINA STATE LIBRARY
1500 Senate St.
Columbia   29211-1469
(803) 734-8666

GREENVILLE COUNTY LIBRARY
SYSTEM
25 Heritage Green Place
Greenville 29601-2034
(864) 242-5000

SOUTH DAKOTA

SOUTH DAKOTA STATE LIBRARY
800 Governors Dr.
Pierre   57501-2294
(605) 773-3131
(800) 592-1841 (SD residents)

DAKOTA STATE UNIVERSITY
Nonprofit Management Institute
Nonprofit Grants Assistance
820 N. Washington
Madison   57042
(605) 367-5382

BLACK HILLS STATE UNIVERSITY
E.Y. Berry Library-Learning Center
1200 University St. Unit 9676
Spearfish 57799-9676
(605) 642-6834

TENNESSEE

UNITED WAY OF GREATER
CHATTANOOGA
Center for Nonprofits
630 Market St.
Chattanooga 37402
(423) 265-0514

KNOX COUNTY PUBLIC LIBRARY
500 W. Church Ave.
Knoxville   37902
(865) 215-8751

MEMPHIS & SHELBY COUNTY
PUBLIC LIBRARY
3030 Poplar Ave.
Memphis 38111
(901) 415-2734

NASHVILLE PUBLIC LIBRARY
615 Church St.
Nashville   37219
(615) 862-5800

TEXAS

AMARILLO AREA FOUNDATION
Grants Center
801 S. Filmore, Suite 700
Amarillo 79101
(806) 376-4521

HOGG FOUNDATION FOR
MENTAL HEALTH
Regional Foundation Library
3001 Lake Austin Blvd., Suite 400
Austin 78703
(512) 471-5041

BEAUMONT PUBLIC LIBRARY
801 Pearl St.
Beaumont   77704-3827
(409) 838-6606

CORPUS CHRISTI PUBLIC LIBRARY
Funding Information Center
805 Comanche St.
Reference Dept.
Corpus Christi   78401
(361) 880-7000

DALLAS PUBLIC LIBRARY
Urban Information
1515 Young St.
Dallas   75201
(214) 670-1487

SOUTHWEST BORDER NONPROFIT
RESOURCE CENTER
1201 W. University Dr.
Edinburgh 78539-2999
(956) 384-5920

UNIVERSITY OF TEXAS AT EL PASO
Institute for Community-Based
Teaching and Learning Community
Non-profit Grant Library
500 W. University, Benedict Hall, Rm.
103
El Paso 79968-0547
(915) 747-7969

FUNDING INFORMATION CENTER
OF FORT WORTH
329 S. Henderson St.
Ft. Worth 76104
(817) 334-0228

HOUSTON PUBLIC LIBRARY
Bibliographic Information Center
500 McKinney Ave.
Houston 77002
(832) 393-1313

NONPROFIT MANAGEMENT AND
VOLUNTEER CENTER
Laredo Public Library
1120 E. Calton Rd.
Laredo 78041
(956) 795-2400

LONGVIEW PUBLIC LIBRARY
222 W. Cotton St.
Longview 75601
(903) 237-1350

LUBBOCK AREA FOUNDATION, INC.
1655 Main St., Suite 209
Lubbock 79401
(806) 762-8061

NONPROFIT RESOURCE CENTER
OF TEXAS
7404 Hwy. 90 W.
San Antonio 78212-8270
(210) 227-4333

WACO-MCLENNAN COUNTY
LIBRARY
1717 Austin Ave.
Waco 76701
(254) 750-5941

NONPROFIT MANAGEMENT
CENTER OF WICHITA FALLS
2301 Kell Blvd., Suite 218
Wichita Falls 76308
(940) 322-4962

UTAH
GRAND COUNTY PUBLIC LIBRARY
25 South 100 East
Moab 84532
(435) 259-5421

SALT LAKE CITY PUBLIC LIBRARY
210 E. 400 S.
Salt Lake City 84111
(801) 524-8200

VERMONT
ILSLEY PUBLIC LIBRARY
75 Main St.
Middlebury 05753
(802) 388-4095

VERMONT DEPT. OF LIBRARIES
Reference & Law Info. Services
109 State St.
Montpelier 05609
(802) 828-3261

VIRGINIA
WASHINGTON COUNTY
PUBLIC LIBRARY
205 Oak Hill St.
Abingdon 24210
(276) 676-6222

HAMPTON PUBLIC LIBRARY
4207 Victoria Blvd.
Hampton 23669
(757) 727-1314

RICHMOND PUBLIC LIBRARY
Business, Science & Technology Dept.
101 E. Franklin St.
Richmond 23219
(804) 646-7223

ROANOKE CITY PUBLIC
LIBRARY SYSTEM
Main Library
706 S. Jefferson
Roanoke 24016
(540) 853-2471

WASHINGTON
MID-COLUMBIA LIBRARY
1620 South Union St.
Kennewick 99338
(509) 783-7878

KING COUNTY LIBRARY SYSTEM
Redmond Regional Library
15990 NE 85th
Redmond 98052
(425) 885-1861

SEATTLE PUBLIC LIBRARY
Fundraising Resource Center
1000 4th Ave.
Seattle 98104
(206) 386-4636

SPOKANE PUBLIC LIBRARY
Funding Information Center
906 W. Main Ave.
Spokane 99201
(509) 444-5300

UNIVERSITY OF WASHINGTON
TACOMA LIBRARY
1900 Commerce St.
Tacoma 98402
(253) 692-4440

WEST VIRGINIA
KANAWHA COUNTY PUBLIC
LIBRARY
123 Capitol St.
Charleston 25301
(304) 343-4646

WEST VIRGINIA UNIVERSITY AT
PARKERSBURG LIBRARY
300 Campus Dr.
Parkersburg 26101
(304) 424-8260

SHEPHERD UNIVERSITY
Ruth A. Scarborough Library
King St.
Shepherdstown 25443-3210
(304) 876-5420

WISCONSIN
UNIVERSITY OF
WISCONSIN-MADISON
Memorial Library, Grants Information
Center
728 State St.
Madison 53706
(608) 262-3242

MARQUETTE UNIVERSITY
MEMORIAL LIBRARY
Funding Information Center
1355 W. Wisconsin Ave.
Milwaukee 53201-3141
(414) 288-1515

UNIVERSITY OF WISCONSIN—
STEVENS POINT
Library—Foundation Collection
900 Reserve St.
Stevens Point 54481-3897
(715) 346-2540

WYOMING
LARAMIE COUNTY COMMUNITY
COLLEGE
Instructional Resource Center
1400 E. College Dr.
Cheyenne 82007-3299
(307) 778-1206

CAMPBELL COUNTY
PUBLIC LIBRARY
2101 4-J Rd.
Gillette 82718
(307) 687-0115

TETON COUNTY LIBRARY
125 Virginian Ln.
Jackson 83001
(307) 733-2164

SHERIDAN COUNTY FULMER
PUBLIC LIBRARY
335 West Alger St.
Sheridan 82801
(307) 674-8585

PUERTO RICO
UNIVERSIDAD DEL SAGRADO
CORAZON
M.M.T. Guevara Library
Santurce 00914
(787) 728-1515

Participants in the Foundation Center's Cooperating Collections network are libraries or nonprofit information centers that provide fundraising information and other funding-related technical assistance in their communities. Cooperating Collections agree to provide free public access to a basic collection of Foundation Center resources during a regular schedule of hours, along with free funding research guidance to all visitors. Many also provide a variety of services for local nonprofit organizations, using staff or volunteers to prepare special materials, organize workshops, or conduct orientations.

A key initiative of the Foundation Center is to reach under-resourced and underserved populations throughout the United States who are in need of useful information and training to become successful grantseekers. One of the ways we intend to accomplish this goal is by designating new Cooperating Collection libraries in regions that have the ability to serve the nonprofit communities most in need of Foundation Center resources. We are seeking proposals from qualified institutions (i.e., public, academic, or special libraries) that can help us carry out this important initiative. If you are interested in establishing a funding information library in your area, or would like to learn more about the program, please contact Erika Wittlieb, Coordinator of Cooperating Collections, The Foundation Center, 79 Fifth Avenue, New York, NY 10003 (E-mail: eaw@fdncenter.org).

APPENDIX D

Resources/Bibliography

Atienza, J., et al. *Foundation Giving Trends: Update on Funding Priorities.* New York: Foundation Center (annual).

Cambridge Associates. *Sustainable Payout for Foundations: A Study Commisioned by the Council of Michigan Foundations.* Grand Haven, MI: Council of Michigan Foundations, 2000.

Council on Foundations. *Foundation Management Report.* Washington, DC: Council on Foundations (biennial).

DeMarche Associates, Inc. *Spending Policies and Investment Planning for Foundations: A Structure for Determining a Foundation's Asset Mix.* Washington, DC: Council on Foundations, 1999.

Edie, J. *First Steps in Starting a Foundation,* 5th edition. Washington, DC: Council on Foundations, 2002.

Foundation Center. *Corporate Foundation Profiles.* New York: Foundation Center (biennial).

Foundation Center. *Directory of Missouri Grantmakers.* New York: Foundation Center, 2003.

Foundation Center. *The Foundation Directory.* New York: Foundation Center (annual).

Foundation Center. *The Foundation Directory, Part 2.* New York: Foundation Center (annual).

Foundation Center. *Foundation Fundamentals.* 5th edition. Ed. by K. Schladweiler. New York: Foundation Center, 2004.

Foundation Center. *The Foundation 1000.* New York: Foundation Center (annual).

Foundation Center. *Grant Guides.* New York: Foundation Center (annual).

Foundation Center. *Guide to Greater Washington, D.C., Grantmakers on CD-ROM.* New York: Foundation Center, 2004.

Foundation Center. *Guide to Ohio Grantmakers on CD-ROM.* New York: Foundation Center, 2005.

Foundation Center. *Guide to U.S. Foundations, Their Trustees, Officers, and Donors.* New York: Foundation Center (annual).

Foundation Center. *National Directory of Corporate Giving.* New York: Foundation Center (annual).

Foundation Center. *The PRI Directory: Charitable Loans and Other Program-Related Investments by Foundations.* New York: Foundation Center, 2003.

Freeman, D., J. Edie, J.Nober, and the Council on Foundations. *The Handbook of Private Foundations.* Washington, DC: Council on Foundations, 2005.

Giving USA Foundation. *Giving USA.* Ed. by M. Brown. Indianapolis, IN: Giving USA Foundation (annual).

Gray, B. *Health Policy Research and Foundation Grantmaking: Identifying Areas Where Foundations Fill the Gap.* New York: Foundation Center, 2004.

Lawrence, S. *Children and Youth Funding Update.* New York: Foundation Center, 2002.

Lawrence, S. *Family Foundations: A Profile of Funders and Trends.* New York: Foundation Center, 2000.

Lawrence, S. *Foundation Funding for Children's Health: An Overview of Recent Trends.* New York: Foundation Center, 2005.

Lawrence, S. *Key Facts on Family Foundations.* New York: Foundation Center, 2005.

Lawrence, S. *Update on Foundation Health Policy Grantmaking.* New York: Foundation Center, 2004.

Lawrence S., and L. Marino. *New York Metropolitan Area Foundations: A Profile of the Grantmaking Community.* New York: Foundation Center, 2002.

Lawrence, S., and L. Marino. *Update on Funding for Higher and Graduate Educational Institutions.* New York: Foundation Center, 2003.

Lawrence, S., J. Atienza, and A. Barve. *Foundation Yearbook: Facts and Figures on Private and Community Foundations.* New York: Foundation Center (annual).

Lawrence, S., R. Gluck, and D. Ganguly. *California Foundations: A Profile of the State's Grantmaking Community.* New York: Foundation Center, 2001.

Muirhead, S. *Corporate Contributions.* New York: Conference Board (annual).

Renz, L. *Assessing the Post-9/11 Funding Environment: Grantmakers' Perspectives.* New York: Foundation Center, 2002.

Renz, L. *Foundations Today: Current Facts and Figures on Private and Community Foundations.* New York: Foundation Center, 1988–1990.

Renz, L., J. Atienza, H. Seidler, and S. Lawrence. *International Grantmaking III: An Update on U.S. Foundation Trends.* New York: Foundation Center, 2004.

Renz, L., and S. Lawrence. *Arts Funding IV: An Update on Foundation Trends.* New York: Foundation Center, 2003.

Renz, L., and S. Lawrence. "Encouraging a Caring Society Through the Promotion of Volunteerism and Philanthropy: The Role of Grantmaking Foundations." Paper presented at INDEPENDENT SECTOR Research Forum, San Antonio, March 1993.

Renz, L., and S. Lawrence. *Foundation Funding for the Humanities: An Overview of Current and Historical Trends.* New York: Foundation Center, 2004.

Renz, L., and S. Lawrence. *Foundation Growth and Giving Estimates.* New York: Foundation Center (annual).

Renz, L., and S. Lawrence. "A Primer on Foundation Science Support." *Science,* 257, Sept. 18, 1992.

Renz, L., S. Lawrence, C. Camposeco, and J. Kendzior. *Southeastern Foundations II: A Profile of the Region's Grantmaking Community.* New York: Foundation Center, 1999.

Renz, L., and L. Marino. *"Giving in the Aftermath of 9/11: Final Update on the Foundation and Corporate Response,"* in *September 11: The Philanthropic Response.* Ed. by R. Schoff. New York: Foundation Center, 2004.

Renz, L., et al. *Foundation Giving: Yearbook of Facts and Figures on Private, Corporate and Community Foundations.* New York: Foundation Center, 1991–1999.

Salamon, L. *America's Nonprofit Sector: A Primer.* New York: Foundation Center, 1999.

Salamon, L. (ed.) *The State of Nonprofit America.* Washington, DC: Brookings Institute Press, 2002.

Weitzman, M., et al. *The New Nonprofit Almanac and Desk Reference: The Essential Facts and Figures for Managers, Researchers, and Volunteers.* Washington, DC: Independent Sector, 2002.